THE RADICAL
ECONOMIC
WORLD VIEW

The Ideal Worlds of Economics

THE RADICAL ECONOMIC WORLD VIEW

BENJAMIN WARD

Basic Books, Inc., Publishers   New York

Library of Congress Cataloging in Publication Data

Ward, Benjamin N
 The ideal worlds of economics.

 Includes bibliographies and index.
 CONTENTS: book 1.   The liberal economic world view.
—book 2.   The radical economic world view.—book 3.
The conservative economic world view.
   1.   Comparative economics.   2.   Liberalism.
3.   Marxian economics.   4.   Conservatism.   I.   Title.
HB90.W37   330   78–54497
ISBN: 0–465–03199–4
ISBN: 0–465–03926–X (v. 1) pbk.
ISBN: 0–465–06818–9 (v. 2) pbk.
ISBN: 0–465–01396–1 (v. 3) pbk.

Contents

PART I

The Optimal Radical Economic World View

1. Introduction   3
2. Contemporary Capitalism: The Charges   8
3. The Structure and Tendency of Societies   15
4. Monopoly Capitalism: Structure and Tendencies   24
5. Exploitation Under Monopoly Capitalism   33
6. Government and Monopoly Capital   42
7. Instability and Crisis   50
8. Development and Imperialism   57
9. The Rise of Socialism: The Soviet Union   66
10. The Rise of Socialism: Yugoslavia and China   74
11. Transitions   82
12. The Future   90

PART II

Commentary

13. Radical World View and Radical Economics   97
14. Baran and Sweezy   102
15. Stagnation and Surplus Absorption   112
16. Alienation   116
17. Horvat   121
18. Technical Economics vs. Radical Economics?   127
19. The Role of Revolution   132

NOTES   135
SUGGESTIONS FOR FURTHER READING   147
INDEX   149

PART I

The Optimal Radical Economic World View

ACKNOWLEDGMENTS

Someone who, like myself, came back to the serious study of Marxist interpretations of contemporary capitalism in the sixties is bound to have been strongly influenced by younger students of the subject. Of the many who have taken time out from other activities to continue my education, I would particularly like to thank George Evans, Bob Harris, David Kotz, Michael Reich, Richard Roehl, John Roemer, and Alan Shelly. A considerable number of dissertations have been completed in recent years at Berkeley that deserve broad circulation among students of the political economy of contemporary capitalism, and which inform the present work. These include Marilyn Goldberg's study of the sharing of housework, Louis Green's study of the economics of establishing a separate black state in the United States, David Kotz's study of the influence of finance capital on industrial corporations, John Roemer's study of Japanese-American competition in other countries, Sam Rosenberg's study of labor market duality, Don Shakow's study of economy-based political power in the Soviet Union in the twenties, Alan Shelly's study of the macroeconomics of depression, and John Willoughby's study of the relation between trade structure and imperialism. Taken as a group, these works demonstrate the viability of radical orientations in supporting scholarly interpretation of the major economic problems of our time.

CHAPTER 1

Introduction

THERE ARE two basic defining qualities of a radical. The first is a commitment to the wretched of the earth, and with it a recognition of their humanity, their dignity, their rights. Sometimes the commitment is acquired abstractly by reading and thinking and in discussion. But attitudes formed in this way have little real force until they are grounded in experience, until in some social sense one has become a part of this large and still growing community of the dispossessed. But, of course, the great majority of those possessing a commitment to the wretched of the earth acquired it by the simple act of being born into membership.

That commitment can make a social worker or even a priest as well as a radical. To become a radical, a second quality is also necessary. This is a firm belief that the wretchedness of much of humanity is unnecessary and that it cannot be eliminated within the framework of existing society. This notion can be acquired intuitively by almost any observant person living in capitalist society. There are those constant contrasts between extremes of deprivation and wealth, and that peculiar phenomenon of scarcity in the midst of plenty, of idle hands together with shortages.

But unless this insight into the basic insanity of capitalist institutions is buttressed by a firm grounding in radical thought, it may not survive. Radicals who do not possess such knowledge are subject to rather violent swings in orientation in the face of changing times, and particularly of adversity. For there are too many comfortable theories around explaining why things have to be as they are, and there are times when it is very convenient to accept one of them. Theoretical grounding is an essential defense against this. But, of course, abstract understanding plays a more substantive role too. It is out of a radical understanding as to how the world works that one acquires a successful strategy for changing the world. Radical thought is not an object of esthetic beauty, but it is a fundamental tool in the service of these two aims.

Radical political economy is at the heart of radical thought because material deprivation is at the heart of most of the misery in the world today. Alienation need not be accompanied by material deprivation; however, the sense of meaninglessness typically grows out of the inhumane aspects of both production and consumption activities, and of the dehumanizing effect on affluent and poor alike of living with such continuing misery. The central task of radical thought is to understand these problems and to understand what to do about them.

Essentially, radical political economy offers a general view of how the world works in the economic and related spheres of human action. This view must simultaneously vindicate the defining qualities of a radical, be consistent with the known facts, be persuasive, and provide an agenda for action. It is not a branch of technical economics, though it draws on research results, from whatever area, that are useful for its purposes. Basically, it tells the big story of how the world works, but in such a way that the little stories of how countries, classes, and even individuals function can be fitted in. It has been in process of development for two centuries, at least since the French Revolution, but its main concern must be to understand the contemporary world, that being the only one we can change.

At the moment of writing (1978), radical political fortunes in many parts of the world of monopoly capitalism are at a relatively low ebb; this is particularly true of the United States. Also, there have been recent setbacks in some developing countries, most notably perhaps in Bolivia and Chile. This sort of thing has happened before, and it is important to understand why it happens, why it will be followed by an upsurge, and why each upsurge tends to surge up farther than the one before. Today hope, confidence in the victory of socialism, lies in three facts. First, there is the fact of the existence of a number of socialist countries who have established their ability to defend themselves successfully against capitalist depredations without destroying the essence of socialism within their societies. Second, there is the fact of the size and power of socialist movements around the world. We are living at a time of very high-low tides, one that is accompanied by surgent socialist forces in many parts of Africa, the Arab world, southern Europe, and East Asia. The radical base is growing larger and even though in the best of circumstances not all of these movements will bear immediate fruit, even the superficial trends are favorable.

Third, and not to be undervalued, there is the fact of the upsurge in radical economic thought. Radical intellectual fortunes have had their ups and downs in the twentieth century, just as radical political fortunes have. However, the causal factors are somewhat different. Obviously they are connected; for example, political oppression of radical intellectuals and the constraints imposed on radical intellectual dialogue by the demands of political action have taken their toll. But there are important underlying factors that have affected the course of twentieth-century radical thought. Most important of these is the history of monopoly capital itself. Created, roughly speaking, around the turn of the century out of capitalism's first major survival crisis and itself bringing great disturbance to the world, it has taken a long time for monopoly capital to reveal its fundamental properties to the radical analyst. However, by the mid-fifties enough evidence was in, and several broad-gauge appraisals of contemporary society from the perspective of the radical economist were published. The first of these, Paul Baran's *Political Economy of Growth*,[1] contained a general theory of the tendencies in modern capitalism, with emphasis on the interaction between developed monopoly capitalism and the Third World economies. Later Baran collaborated with Paul Sweezy in producing *Monopoly Capital*,[2] a survey of the processes of waste and exploitation in contemporary America and of their implications. A Belgian Marxist, Ernest Mandel, in his *Marxist Economic Theory*,[3] provided a general historical analysis of world economic development in the nineteenth and twentieth centuries, based both on Marxian theory and on recent empirical research. He has recently updated some of these views in a large work, *Late Capitalism*.[4] And Yugoslav economist Branko Horvat in his *Toward a Theory of Planned Economy*[5] offered a theory of the underlying processes of change in the contemporary world with emphasis on the emergence of participation as a central part of recent social-economic development.[6]

These works demonstrate the power of radical analysis in exposing fundamental processes in contemporary society; they also demonstrate the viability of Marxism as a basic tool of economic analysis. But, of course, such works are synthetic in nature. They cannot provide a detailed picture of the problems we face today or offer a detailed analysis of proposed solutions. Work of this kind can now be found in considerable abundance in a variety of journals that deal with radical economic analysis, such as the *Review of Radical Political Economy*, published in Ann Arbor, *Socialist Revolution*, published in San Francisco, *Monthly Review*, New York, *New Left Review*, London, and *Cambridge Journal of Economics*, Cambridge, England, to mention only the most interesting of the Anglo-American journals. And so, for the first time in half a century, a rich new intellectual base has been emerging, one on which further upsurges in understanding can be built.

Another central factor in the resurgence of radical economic thought came from the policies of socialist countries and their defense by socialist leaders. Their insights into the nature of society are an amalgam of their practical experience in organizing both revolution and socialist society, and of their intellectual experience in assimilating the major works of radical thought. Though their writings are typically not scholarly in quality, these leaders' works are a major source of insight into the problems posed by the transition to socialism and the establishment and stabilization of socialist momentum in postrevolutionary society. At least one of them, Mao Tse-tung, has fundamentally redirected the radical analysis of the process and needs of economic development.

The socialist countries, of course, publish a great variety of relevant materials for radical economics. Perhaps of most interest in the context of this chapter are the general works appraising contemporary world economy from the perspective of the respective leaderships of these societies. For example, the Soviet Union's Communist Party publishes an English-language version of their *Political Economy*, which gives special weight to the role of the Soviet Union in shaping twentieth-century experience. Chinese views are captured in their political economies of capitalism and socialism, one of which has recently appeared in English translation,[7] and especially in the works of Mao. His *Selected Works* are available in a cheap five-volume English language edition.[8] Horvat, mentioned above, offers a perspective on Yugoslavia, and some important aspects of the Cuban revolution are to be found among the speeches of Ché Guevara, collected by John Gerassi in *Venceremos*.[9]

A third source of resurgence in radical economic thought has been the radical youth movement of the sixties. Of course, the forte of activists within that movement was not so much radical economic analysis as it was a well-articulated awareness of the contradictions of monopoly capitalist society. As a consequence of their work, many millions of people who do not think of themselves as radicals have had permanently etched on their minds the great gap between the civics textbook ideals of a liberal society and the facts of liberal action in dealing with such things as racism, sexism, inequality, and war.

The undogmatic nature of this movement had its effect on radical thought. The distinction between the Old Left and the New Left, very striking, for example, in the United States in the early sixties, has essentially disappeared with the changing circumstances of economic and political life, and with a recognition of fundamentally common goals. An environment has thus been established in which radical dialogue on central issues of radical thought can once again take place productively.

These factors in combination have generated an unprecedented interest in radical economics. Not only are there a large number of young men and women seriously engaged in the effort of developing a powerful radical instrument of economic analysis, but the intellectual quality that they bring to bear on that effort is extremely high, far higher, I believe, than in any previous generation. One can reasonably expect that the next decade will bring a revolutionary improvement in the breadth and depth of radical economic thought.

But that is not to suggest that consensus will be reached on every major point. At present a genuine dialogue does not yet exist between the radical economists of most socialist countries and their Western counterparts, or even among economists in different socialist countries. The work of Third World and Western economists is only in an early stage of mutual interaction. And among Western radicals there is a great variety of views; some of the few remarks that have so far been made in this introduction would certainly be disputed.

This situation suuggests the opportunities that exist for further development of radical thought. It also says something about the dialectical nature of current radical political economy. A serious student of radical political economy will have an overview. She [10] considers this overview to be optimal in the sense that it characterizes better than the radical alternatives the ways in which the contemporary world works. But such a student must also recognize that many of those alternative views are not to be dismissed from further consideration. Rather, they are a part of radical political economy, the subject matter of the dialogues that, in combination with further study and experience, will resolve portions of the disagreements. Thought is as dialectical as life; contradictions must be recognized and built upon. The views expressed in this work are, hopefully, one side of a discussion, and will no doubt require revision in the light of even initial responses.

CHAPTER 2

Contemporary Capitalism: The Charges

CONTEMPORARY MONOPOLY CAPITALISM has a number of achievements to its credit. For example, putting a man on the moon was a very neat trick. No one would deny the great effort and skill that went into the performance of that trick. However, perhaps a majority of Americans have felt that this effort was undertaken only because of a mistaken notion of appropriate social priorities. A radical, on the other hand, would be likely to claim that it was no mistake but was rather a typical manifestation of the peculiar inhumanity of monopoly capitalism, in which it was decided to leave millions of American citizens hungry, ill housed, their massive health problems largely untreated, while performing this little $20 billion feat of technical wizardry.

This example suggests why in the present chapter no attempt is made to offer a "balanced" picture of the achievements and problems of contemporary capitalism. Instead, we in effect bring the defendant to the bar and detail her crimes, in order to determine whether these crimes are of sufficient magnitude that he be deemed a menace to public safety. The crimes are not mitigated by the achievements because even they are all tainted by this distorted sense of values. Putting a man on the moon is of no relevance to the case, not because there is no intrinsic interest in the episode, but because it was bought at the expense of a continuation of capitalism's failures. Its significance is to be measured by those failures.[1]

Inequality

The defendant is unlikely to contest this charge.[2] In absolute terms it refers to the 20 to 30 million Americans who fall below the poverty line and whose lives clearly are rendered nightmarish by the various forms of material deprivation they must endure. The official American poverty line was drawn with a political aim of "ensuring that very few of those who were listed as living in poverty would not in fact be," [3] so that the true figures probably are much higher and have been rising alarmingly in recent years. Furthermore, the line captures the basis of material deprivation in a complex modern society only in a very crude way, so that it is quite possible that as many as 50 million Americans find their own and their children's lives stunted by serious deprivation, most likely of adequate health care or housing. And all this in the world's richest country.

But this is only the beginning. Recent estimates put the number of the world's citizens who suffer serious effects from hunger and malnutrition at well over half a billion, with the upper limit possibilities getting close to a billion.[4] And these figures, it should be emphasized, refer to serious malnutrition, the kind that stunts growth, creates mental retardation, and generates susceptibility to the many diseases that shorten life dramatically. Furthermore, this hunger is not a product of the inadequacy of the world's food supply. It is rather a matter of its distribution, which is a social property of the problem, determined by the kind of institutions that govern economic life. In this case, of course, the institution is capitalism.

The reason for associating capitalism with these problems can be suggested by looking at the other end of the spectrum in American society. Here we find 1 percent of the populace owning well over a quarter of the privately held wealth, over two thirds of the corporate stock in private hands, and almost all the municipal bonds. Here we find over five hundred citizens reporting an annual *income* of over a million dollars. The very rich tend to have inherited much of their wealth and to hire skilled professional administrators to increase it for them and to reduce their tax liability. The results suggest that under capitalism the distribution rule is the familiar one: Those who have, get. A large segment of the poor—a fifth or so—have negative wealth; that is, they owe more than they own.

In the "developing" capitalist countries, the situation is generally much worse. The top 5 percent tend to have up to twice the proportion of national income as in the developed countries.[5] Presumably the wealth distribution follows the income distribution too, though it is much more highly skewed. Private ownership of the means of production is clearly the key to these appalling figures. Those little pieces of paper that give a citizen the legal right to some portion of a firm or government's income are what generate the extremes of opulence and misery.

Inequality of wealth and income provides the basis for reproducing the social structure of a society generation after generation by creating very great inequality of opportunity. The children of the affluent are better eduucated in better schools and, through parental influence, are likely to be given preference in entry to privileged jobs, such as the professions. Children of the poor suffer the handicaps of material deprivation compounded by the unequal treatment they receive in the marketplace. Once again, at the heart of the inequality of opportunity is the inequality of wealth, and at the heart of the inequality of wealth is the system of private ownership of the means of production.

Imperialism

Imperialism is another charge where the evidence is decisive.[6] In the first place, there are a number of overt acts that cannot be concealed. In a long list of countries, including Chile, Brazil, Vietnam, the Dominican Republic, Lebanon, and many others, the United States has assisted in the generation of a regime favorable to its interests. In some cases this has meant sending in the marines (Lebanon, Dominican Republic), in some a systematic financial boycott (Chile), in other cases various forms of clandestine assistance to fascists (Greece), in some massive military intervention (Korea, Vietnam), and in almost all cases success of the political venture is followed by opening American coffers, in the form of aid and loans, to the new anticommunist, procapitalist government.

What is all this in aid of? One thing does stand out—the tendency for a large increase in American foreign investment to follow closely upon the success of such ventures. This investment is made by private American business, mostly large multinational corporations, and is done for a single reason, namely, to make profits. The association between intervention and investment is strong enough to suggest, even without any theory, that there is some sort of a causal connection.

Second, of course, there is anticommunism. That this is a distinct aspect of American foreign policy there can be little doubt. Even if one cannot quite take seriously the speeches innumerable influential public figures make in support of this policy line, there is the more direct evidence of the connection between military and economic aid and the containment of communism. American's vast giveaway programs were mostly spent around the periphery of the communist world and were aimed at sustaining appropriately reactionary and procapitalist governments in power as a buffer against any further communist expansion. Yugoslavia, a major aid recipient, was an exception to the rule but not to its spirit. More typical were the suppression of revolution in Greece, the massive support for Chiang's Taiwan, and the far more massive investment in General Thieu and South Vietnam. This buffer, as long as it lasts, provides a barrier around the foreign investment operations that have proven so profitable to monopoly capital.

The consequences of this sort of behavior for the ordinary people of

the world are very great. Militarization diverts over $200 *billion*
worth of resources a year to the sterility of the arms race and indoc-
trinates millions in the arts—and the practice—of violence.[7] Industrial
development is oriented toward profits, which automatically precludes
doing anything about the misery of the poor, who, of course, cannot pay
a profitable price for the things they need. Social revolutions are
suppressed, or prolonged, or distorted, by the appalling scale of the
violence and oppression engendered by the worldwide capitalist system
of economic organization.

Racism and Sexism

Racism and sexism, of course, antedate capitalism. The charge here is
that capitalist institutions have tended to support and enhance existing
forms of discrimination. One evidence of this lies in the wage data,
which show blacks and women obtaining less than two thirds the pay
of men. In some capitalist countries the figure drops below one half.
There is often a substantial direct discrimination discount for people
doing the same work, but most of the difference reflects the low skill
and temporary nature of the jobs into which the discriminated groups
tend to be forced.[8]

Beyond this direct material effect of discrimination lies the subtler
generation of inequality by means of stereotyping, by establishing in
young children's minds low upper limits to their prospects if their sex or
skin color differs from the "norm" of the white male. Added to this are a
variety of legal and quasi-legal disabilities, which control black access
to middle-class housing, female access to abortion, and the like. Among
the nastiest of these is the treatment accorded blacks who are caught up
in the criminal and penal processes of American society.

Behind racist and sexist behaviors lies the phenomenon of class.
This is reflected in the high incidence of black- and women-headed
families who show up in the lowest ranks of the income distribution.
It is reflected in the low-wage history of the South, where racism has
been used as a most effective device to keep wages of both whites and
blacks down and to prevent effective unionization. And it is reflected in
intergenerational mobility, where poverty is shown to breed poverty in
the next generation via the effects of material deprivation and stereotyping.

Alienation

The settlers' town is a strongly built town, all made of stone and steel.
It is a brightly lit town; the streets are covered with asphalt, and the garbage
cans swallow all the leavings, unseen, unknown and hardly thought about.

The settler's feet are never visible, except perhaps in the sea; but there you're never close enough to see them. His feet are protected by strong shoes although the streets of his town are clean and even, with no holes or stones. The settler's town is a well-fed town, an easygoing town; its belly is always full of good things. The settler's town is a town of white people, of foreigners.

The town belonging to the colonized people, or at least the native town, the negro village, the medina, the reservation, is a place of ill fame, peopled by men of evil repute. They are born there, it matters little where or how; they die there, it matters not where, nor how. It is a world without spaciousness; men live there on top of each other, and their huts are built one on top of the other. The native town is a hungry town, starved of bread, of meat, of shoes, of coal, of light. The native town is a crouching village, a town on its knees, a town wallowing in the mire. It is a town of niggers and dirty arabs. The look that the native turns on the settler's town is a look of lust, a look of envy; it expresses his dreams of possession—all manner of possession: to sit at the settler's table, to sleep in the settler's bed, with his wife if possible. The colonized man is an envious man. And this the settler knows very well; when their glances meet he ascertains bitterly, always on the defensive, "They want to take our place." It is true, for there is no native who does not dream at least once a day of setting himself up in the settler's place.[9]

These two paragraphs are a black French psychologist's account of the effect of colonialism on a human community. It describes succinctly the way in which a class society distorts the values of the participants. The oppressed dream of acquiring the status of oppressor, the oppressors strive to continue the oppression, while minimizing their contact with its inhumane consequences. Both end up in an alienated state, unable, within the context of their society, to find a means of expressing themselves through normal human relations.

But, of course, this account is not relevant only for Algeria; it fits the situation in capitalist society everywhere. The ghettos of the poor are to be found all over the United States, the leading capitalist country, as are the attitudes of "settler" and "native." But in the heartland of monopoly capital the alienation is extended in a variety of ways. In large-scale business one may think of the shift from management to administration. That is, the captain of industry shouting out her dictatorial orders to underlings who have no doubt as to who is boss and how powerful she is has become an anachronism. In her place is the almost anonymous administrator, who creates a complex web of carrots and sticks, each one of minor significance, that guides the underlings and serves to shield the individual's awareness of the brutal coercion and oppression that continues to sustain the system.

Or one can think of the subdued pluralism of a one-dimensional, uncritical democracy. In the "democracy" we know in the United States, one can find deeply critical books, but they turn out to be irrelevant in the institutional context of our political system. In the latter a massive effort is undergone to make a choice where there is no choice, to select among candidates whose effects on policy are virtually indistinguishable. Or one can think of the technocratic environment of the worker, whose worklife is dominated, not by his needs as a creative human being, but by the need for certain kinds of manual assistance by the machines a profit-oriented capitalism tends to design.[10]

Or one can think of the vulgar and commercial world of leisure time, in which the fundamental settler-native structure of society presses its members to ever greater paroxysms of consumption in the effort to preserve their hard-won status, whatever that may be. In all these cases alienation, the sense of fundamental meaninglessness, of lack of a strong commitment to some coherent set of values, is the product of living in an environment in which the individual has no feeling of being a participant in making the basic choices that shape her life. The property of that society which generates these various kinds of alienation is its class nature, which is, of course, an inherent property of capitalism.

Irrationality

The irrationality of capitalist society lies in the insane contrasts in which it abounds. There is the contrast between those who are deeply deprived and those who are alienated by the very quantity of goods they possess and the intense activity required to display them to friends and neighbors.. There is the ever-present contrast between the ever-increasing size and "quality" of the armed forces, usually called the instrument of national "defense," and the increasing incidence of war and of the threat of nuclear war. There is the contrast between the increasing need of the growing body of the world's deprived and, under current institutions, helpless peoples, and the increasing lack of manifested concern for other humans. And so on.

And this list deals only with the visible contrasts. Lying behind it is the fundamental contrast between ideals and actions in modern capitalist society. The ideals of mutuality and respect are not dead; they are a part of the inner nature of all the actors in our drama, dope fiends, alcoholics, housewives, tycoons, and all the rest. But they have been suspended by pressures stemming from the institutions under which they all live, under which, as individuals, they are constrained to live.[11]

Conclusion

This brief sketch of the charges against contemporary capitalism indicates the depth and breadth of the failure of its institutions to deal effectively with basic problems of human life. It also indicates the poverty of such measures as output per capita or "real" income, which are typically used by conventional economists to measure progress under capitalism.

The conventional economist reading this list reacts by saying, One thing the charges do not indicate is whether another form of society can do better than this. That is quite true, aside from the feeling one must get that there *has* to be a better way than this just because this is so bad. One way to respond to that objection is to point to one of the obvious and uncontested achievements of socialist societies: They have dramatically reduced inequality of income distribution. And they have done it by taking a very simple step, namely, by placing the means of production under social control and thereby eliminating at a stroke all those little pieces of paper that entitle the most affluent of capitalist citizenry to continue to enrich themselves simply by virtue of possessing the bits of paper. Many of the legion of toilers in the service of the former capitalists, lawyers, tax men, brokers and the like, were put to useful work. And the proceeds from this operation were used largely to provide a basic material cushion for the citizenry, so that hunger, for example, is not a phenomenon to be observed in these societies. That may not be a full answer to the objection, but it goes quite a ways.

Another objection is that the charges in themselves are of little help because they do not establish causation, do not establish the necessary connections between capitalist institutions and these facts. Much of the rest of this book is devoted to filling this gap.

CHAPTER 3

The Structure and Tendency of Societies

KARL MARX has been dead for ninety years; to put it another way, you can look all through her massive collection of writings without finding a single fact about life in the twentieth century. Since he put great emphasis on the flow of history as a basis for social understanding, even a Marxist is bound to feel that there are serious inadequacies in her works as descriptions of contemporary society. But not only has history thrown up new facts, social thinkers have been at work in massive numbers interpreting these facts and unearthing and interpreting new ones about earlier times. Here too Marx's work necessarily displays some inadequacies.[1]

Nevertheless, Marx still has some fundamental things to say to the student of contemporary society. One of these is his emphasis on broad, underlying, long-run forces that condition even the daily life of humans for entire historical eras. Another is her emphasis on changing labor productivity, or the forces of production, as he called it, as creating conditions favorable for new social developments. A third is an analysis of the role of class in conditioning the ways in which new social developments affect the relations of production, that is, the ways in which human beings interact while creating the social product. Yet another is Marx's emphasis on the way in which the upper class of a society is able to use the relations of production as a device for stripping those who create the social product by their work of any excess product beyond their essential consumption. A final factor is his emphasis on a nonlinear dynamics in all societies, that is, a process of change that proceeds at times steadily, at times with increasing instability, at times with cataclysmic change.

This small list by no means exhausts the durable portion of Marx's work, but it does capture the central ideas.[2] In discussing and illustrating them with some selected aspects of historical development up to the eve of the twentieth century, we will not follow Marx's own analysis rigidly. Instead, an attempt will be made to modify the orthodox Marxist picture wherever new facts and analyses suggest that is necessary.[3] But this chapter is not aimed just at illustrating concepts; the historical movements are relevant for understanding contemporary times because today we are in the midst of two of history's most fundamental eras of change: the era of the industrial revolution and the era of the rise of socialism.

The Peasant Revolution

The first great revolution in labor productivity occurred some thousands of years ago. Previously, humans had existed by harvesting the fruits of nature through hunting, gathering, and fishing. The new ingredient was settled agriculture, the planned and deliberate cultivation of the things that were later to be harvested. The result was a potential increase in output per person of manyfold and a fundamental transformation in the relation of people, both to nature and to one another.

Consider for a moment one of its manifestations, namely, the great river valley civilizations of Egypt, Mesopotamia, India, and China. The creation of massive irrigation works led to an increase in output per person of over tenfold as compared with simpler production techniques. This in turn made it possible for the surplus beyond the peasants' own needs to be used to support a large population engaged in other pursuits, from priestly duties to craft production to military activity to city building. Large cities were built to house a substantial class of rulers and their hangers-on and of workers producing goods not essential to human survival.

For all this to take place, some system of distribution had to be created. The details of the system are still unknown, but roughly the palace or temple obtained the right, backed up, of course, by force, to a share of the peasants' output. In return for this the peasant was presumably given some assurance of participation in afterlife existence and was given some protection from famine by the palace storage facilities. At any rate it does not seem that the peasant family itself retained any significant portion of the surplus it created.

The forms and the stability of life varied considerably from one such civilization to another. Egypt enjoyed extraordinary stability—it was perhaps the most stable society known to history. In southern Mesopotamia, on the other hand, such things as the high water table and the nearness of the sea meant that irrigation canals silted up and land became saline and unusable after three or four generations, bringing about the periodic decline of one city and the rise of another. Also the modest size of cities on the Mesopotamian plain seemed to prevent any one city from maintaining a durable control over the others, so that Mesopotamia's ancient history is an extraordinarily bloody one.

In Mesopotamia and elsewhere one sees some almost cyclical structural movements, an alternation between periods of imperial hegemony and feudal dispersion. This may have been related to the limits to effective control that could be exercised, given the military technology of the day, and the tendency for the best military leaders to be raised and trained near the borders of an empire after a generation or two, thus creating a favorable environment for a revolt. These great cycles of the rise and decline of kings and emperors provide the stuff of political history. They also dramatically affect the lives and well-being of the peasantry that probably comprised four fifths to nine tenths of the population, as lands were ravaged, soldiers conscripted, captured, and enslaved, women and children carried away into slavery, and the like. But they had no great effect on the basic organization of society for a period of two to three thousand years, the same processes of production and surplus extraction providing a stable and durable structure even to the rapidly changing southern Mesopotamian political-geographic scene.

Greek and Roman societies represent a later and somewhat "modernized" version of the river valley civilizations. The status of the peasant in Mesopotamia and Egypt is clearly that of a human being under substantial coercive restraint; however, it appears that her status is not that of a slave, in the sense that he could not be bought and sold at the whim of his master. Rather, she seems to have been bound to the land and to the performing of certain duties outside his own plot of land, including probably the building of irrigation works and possibly the building of temples and tombs. Roman and Greek slaves are "slaves" in the sense made familiar to us from, for example, the history of the American South, where elements of markets were well enough developed for a money price to be placeable at almost any time on a slave. But statuses and obligations, such as the rights and abilities associated with manumission and the breakup of families, tended to vary widely over time and place. The main element of continuity in these societies lay in the fact that the slaves performed the basic labor of the society, and the ruling class, one way or another, extracted the surplus of the slaves' production and put it to their own uses. These uses too varied widely over time and place, reflecting, among other things, variations in culture, the productivity of the land, and the threat of invasion.

As one moves down through history and across cultures, a great variety of institutional forms for extracting surplus from peasant populations emerges. Serfdom of the Western European kind is one method, probably growing up as a reflection of the relative surplus of land and scarcity of peasants in economies where some market relations have come to exist. The nobility is able to control runaways fairly effectively by force, and so uses the fiction that the peasant is "bound" to the land and cannot be bought and sold separately from it or allowed freely to leave it as the device for keeping him at work and getting a portion of her labor product. An alternative form, when labor is relatively less scarce, is share tenancy, in which the land "owner" allows a peasant to work the land and reserves for himself some substantial fraction of the product. Peasant freehold farming has occurred throughout history but as a durable form has been largely restricted to peripheral and less productive areas, such as the mountains, where physical control of the peasant is difficult and the surplus product relatively small.

As one comes closer to modern times, one finds increasing portions of the world's peasant labor force caught up at least marginally in the money economy. This produces some central tendency toward a fairly specific structure of agrarian production relations, often known as debt peonage. In such structures there is a fraction of farmers who own fairly substantial farms and are relatively free of debt and a fraction of farmers who own no land and must either rent land or work as wage laborers on the farms of others. But most peasants lie between these two extremes, owning some land, which is heavily burdened with debt and insufficient for a livelihood, and renting the rest, perhaps combined with some off-farm labor by one or two members of the family. From the point of view of surplus extraction, this form had two key advantages: (1) the ideal of debt-free ownership is held out as a carrot to the farmer and provides a strong motivation to work hard; and (2) debt is an effective instrument for surplus extraction from this hard-working citizenry.

This form of organization has proven to be extraordinarily durable. Of course, a major reason is its efficiency as a resource extractor. The moneylender in less affluent societies often is also a farmer. In China, for example, she would have been either a rich peasant or perhaps a member of the literate and cultured gentry who owned farmland but also had other sources of income. In such more highly developed peasant societies banks can also arise to serve this function, thus transferring the surplus more or less directly out of agriculture.

But other features of this system enhanced its durability. For example, the moneylender was not the only person who could prey on the farmer. Often the middlemen who handled the transfer of the agricultural product to the cities were able to obtain—always in part through the use of coercion—monopoly power over the sale of crops, thus ensuring a minimal price to the actual producer. Yet another feature of this debt-peonage system was that it avoided agglomerations of peasants into large groups who might pose a serious revolutionary threat to the regime. When conditions became intolerable, which happened frequently, there were endless peasant riots and petty revolts. But organization was extremely difficult to achieve on a mass scale among such a dispersed population, and where the material condition and culture of the peasantry might vary quite considerably from one locale to the next. Consequently, a peasantry bound to the production task by this particular set of relations of production has tended to persist almost everywhere except where economies of scale have led to an entirely different mode of farming, or where a conquered people can be effectively bound to a latifundia system of large estates. Debt peonage, a relatively late product of history's first great technical revolution, is still playing a central role in contemporary history.

The Industrial Revolution

The last two centuries have witnessed the rise and spread of world history's second great quantum leap in the productivity of labor. Closely associated with factory production and the use of machines, the first major steps in this social and technical revolution were taken in English cotton and iron and machinery production during the eighteenth century, spread to most of Western Europe, the United States, and Japan during the nineteenth century, and have been reaching out to most of the world's peoples during the twentieth. Once a country gets caught up in this process, an annual average rise of output per person of several percentage points is sustainable over a period of decades, to say the least.

The industrial revolution has some features in common with its predecessor, the peasant revolution. In both cases the change transforms the lives of the overwhelming majority of human beings. In both cases the process has changes in the forces of production, changes in known ways of transforming nature, as an essential precondition. In both cases class differentiation plays a central role in fixing the structures of the new societies, though a wide variety of social forms remain feasible. And in both cases societies that have not made the transition are substantially affected by those that have.

We are still in the midst of this great revolution. Almost all the world's societies are presently changing at unprecedented rates. The longer-run outcomes of the process are still probably not to be discerned with any clarity. Nevertheless, some features of even the earlier phases of the industrial revolution are helpful in understanding the contemporary world.

In the few centuries just preceding the industrial revolution, several factors combined to set it in motion. First there was the development of the world's most effective military technology in Western Europe, combined with the improvement of sea transport to the point at which victorious military force could be transferred to almost any seaside point on the world's surface. The Atlantic nations used this power to loot much of the treasure that great peasant civilizations had accumulated, after which still more treasure was extracted from mines and plantations set up in the conquered areas. This primitive capitalist accumulation was one of the keys to the origin of the industrial revolution.

However, there were equally important domestic developments in Europe. A more stable domestic life had given its usual stimulus to trade. The increased trade, probably assisted by a period of relative scarcity of labor induced by the Black Death and, perhaps, climatic change, and by the influx of monetary metals from the New World, accelerated the movement toward the more modern form of peasant exploitation, debt peonage. This increased mobility and greatly stimulated the development of markets and the money economy, a factor still further enhanced by the rise of the early modern state under the influence of the new military technology. Cities flourished, demand increased, and the greater cultural interactions made many aware of the variety of possibilities in human existence. This heterotic vigor has been a typical feature of such times of relatively intense crosscultural contact. But the creativity, shaped by the emerging new environment, tended to be strongly focused on new ways of making money. These various factors were most strongly combined in post-Civil War England, and so England became the home of the industrial revolution.

A second important aspect of the industrial revolution is the illustration it provides of the strong interdependence between the forces and the relations of production. The industrial revolution is closely associated with the concentration of large numbers of workers in a factory full of machinery. However, in its early phases the new technology did not require such concentration, particularly in textiles and machinery production. It would have been perfectly feasible to continue to use the putting-out system, in which various operations were farmed out by a merchant entrepreneur to workers dispersed in farms and small shops. The rise of the early factory seems to be more a matter of the advantages it offered from the point of view of surplus extraction. Workers could be kept at the job for long hours and their manner and intensity of work subject to close control. The very extensive and extremely brutal use of women and children in these factories was no doubt in part because they were more easily disciplined, though of course they came a good deal cheaper too.

The point of this story is that technology did not determine the outcome. What it did do was open up the possibility of dramatically increasing labor productivity, a possibility that could have been realized under several alternative systems of production relations. The particular one that was chosen was a reflection of the existing relations of production in early modern society, a class society in which an upper class of merchants, petty producers, and putters-out had established, on a small but extensive scale, a market system of exploitation via small-scale production and trade. Their interests dictated the move to large-scale production in factories, and they had the power to implement their desires. The relations of production and the forces of production in combination produced the industrial revolution.

There is yet another, perhaps even more fundamental aspect to the industrial revolution, namely, the line of causation running from the relations of production to the forces of production. Once the factory system was established and had demonstrated the fantastic opportunities for surplus generation that were inherent in it, further efforts at technological change were likely to be slanted strongly toward devices that could be used in factories. The later history of technological change enhancing the productivity of labor is thus in all probability very strongly conditioned by the social environment in which the search was embedded. As time went on the technology clearly became such that in most industries high productivity could only be achieved in large factories. But to the extent that this line of causation was effective, the outcome was by no means strictly necessary. It was a trajectory to some unknown but perhaps very substantial extent determined rather by the social relations of production. The supreme relevance of this point for the analysis of contemporary society needs no emphasis.

As a final contrast between the peasant and industrial revolutions, there is the difference in the distribution of population over the earth's surface. Essentially, the peasant revolution, while permitting an order of magnitude increase in population, permitted only relatively small concentrations of population. Some particularly fertile plains, such as the Nile Valley, acquired large concentrations, but on the whole under 10 percent of the population was concentrated in cities. The industrial revolution, of course, has had just the opposite effect, with the proportion of population engaged in agriculture being reduced below 10 percent in the most advanced countries. The concentration of the exploited class in cities, and in factory agglomerations where they were in close contact with one another, made possible the rise of mass movements of the exploited. The process by which this occurs is a complex one, and there have been all too many ways in which worker organizations have been subverted into tools of the exploiting upper class, but in general terms urbanization of the direct producers has posed a fundamental threat to the stability of the factory system as a method of exploitation.

The Crisis of Nineteenth-Century Capitalism [4]

The unprecedented growth of productive resources and the concomitant rapid social transformations that accompanied the growth have turned the industrial revolution into an era of great social turbulence. The uprooting of traditional modes of life, the throwing of vast numbers of people onto impersonal labor markets, the insecurity of wage labor as a basis for sustaining a family, and the violent swings in the level of economic activity that accompanied the growth trends all contributed to this turbulence. During most of the nineteenth century the growth itself, with the promise of ultimate plenty that it held out, probably played a substantial role in muting social conflict in much of Europe. There were upsurges of revolutionary activity in 1830, again in 1848, and again in 1870, and the level of violent but localized disruption in individual economies waxed and waned throughout the period; but basically nineteenth-century societies undergoing industrial development seemed to retain some tolerable measure of underlying stability.

However, as the nineteenth century drew to a close this situation began to change. Several factors combined to produce a fundamental threat to the newly emergent industrial capitalist system. One of the most important of these was the growing and intensifying rivalry among capitalists. The rivalry took a number of forms. For example, in the

United States it was in considerable measure a consequence of the creation of nationwide markets, which in turn were brought about by the rise of the railroad. Businesses that had been able to maintain local monopolies because of the high transport costs potentially rival factories had to pay suddenly found themselves faced with competition. Often, in a market where several firms operated under negotiated price agreements, these cartellike arrangements tended to break down as the number of sellers to that market increased substantially. In England the process was less intense, partly because of the slower pace of growth, partly perhaps because in a number of areas it was possible to sustain gentlemen's agreements for market sharing and price fixing, and partly because of the relief to local market entry offered by the burgeoning opportunities for foreign investment. The pressure increasingly came from factories in other industrializing countries. Each country had its own unique story to tell in this regard, but intensifying competition was nevertheless a ubiquitous feature of the capitalist scene.

A second factor was economic instability, which seemed to be increasing sharply as the century wore to a close. Business cycles in the anarchic market system are rather like hydrodynamic turbulence: The general lines of causation are understood but a well-developed theory still eludes the theorist, in both cases because of the great variety of factors that can impinge on a given situation. Inventory cycles, construction cycles, cycles of expectations that stimulated overinvestment, financial crises induced by rigidities in the monetary system; all these played their role in generating the business cycle. The intensification of the cycle was related to the accumulated growth, which tended to put a steadily larger amount of productive activity into relatively volatile sectors of the economy, such as investment and the production of other goods whose purchase is easily postponable. The very rapid development of financial markets, an extremely volatile sector in which prayers and promises and chicanery are major "means of production," no doubt played its role as well.

Another major factor in intensifying the cycle was the internationalization of the capitalist system, which was proceeding rapidly during the second half of the nineteenth century. From the eighteen fifties on business cycles in the various industrializing countries became interdependent. In practice this meant that a crisis in one part of capitalism's geography was quickly transmitted to the other parts. Consequently, some potentially stabilizing factors such as foreign investment could not be brought to bear because prospects were simultaneously bad all over.

But the internationalization of capitalism had a far more serious effect than that, for it also generated the era of imperialist rivalry. The economic expansion of capitalism was never separated from the use of violence; the two were linked during the age of primitive capitalist accumulation, and this linkage continued during the period of colonialization of the world. But this deadly competition for raw materials and markets abroad forced capitalists into competing national groups, each with its own fleet and army. This rather substantial conversion of capitalist states into the instruments of national coalitions of capitalists operating internationally created a period of almost annual international crisis. It also led to a very substantial and increasing diversion of resources into the competitive development of the rival fleets and armies.

One more factor needs to be put into the crisis equation, and this the most fundamental of all. In the later nineteenth century labor began everywhere in the developing industrial world to mobilize for economic and political struggle. The factory system was beginning to negate itself as the class consciousness of workers developed their awareness of the basically exploitative structure of the societies in which they lived. Unions were growing everywhere and were becoming more militant. But, even more alarming to the ruling class, workers were also forming political organizations. Social democracy in Germany, the Labour party in England, the Socialist party and the IWW (International Workers of the World) in the United States, and a variety of other political organizations in these countries and elsewhere signified the intensification of the domestic class struggle everywhere.

From the point of view of the stability of capitalism, this struggle could not have occurred at a more unfortunate time. Not only were there the problems of instability, and the threat of war, which would have to be fought with the working classes as cannon fodder, but a sort of economic climacteric had also been reached. The diversion of resources abroad and into military expenditures at home, and perhaps other factors as well, such as the intense rivalry among capitalists, had led to a stagnation in real wages under capitalism. Workers were well aware that labor productivity was continuing its growth but that they were obtaining no share in the increase. Labor turbulence of all forms was increasing at a dramatic pace as capitalism moved through these years and into the early twentieth century.

Such was the scene of crisis. Some sort of explosion seemed clearly imminent, and desperate measures indeed would be required if basically capitalist relations of production were to be preserved by the ruling class. The twisted and brutal adjustments that were made ushered in the era of monopoly capitalism and that bloody, transitional period that we call the twentieth century.

CHAPTER 4

Monopoly Capitalism: Structure and Tendencies

MONOPOLY CAPITALISM is not a rigid structure that was created in the year 1900 and has remained unchanged ever since. Rather, it has come into being piecemeal in response to specific crises and opportunities and, once created, has made numerous adaptations to its structure in the face of newer events. Nevertheless, it will be useful to start the discussion with a description of the principal adaptation to each of the aspects of the first general crisis of capitalism as these were given at the end of the last chapter. Then we will follow monopoly capitalism's course through the century in order to assess its role in the great events of our era and to appraise the tendencies built into it.

Emerging Structure, A Response to Crisis

The first element of crisis discussed in the last chapter had to do with the intensifying rivalries among capitalists.[1] Though the process varied from country to country, the outcome tended to be much the same. There was a strong trend toward substantial concentration, in many cases to the point of monopoly or industrywide cartel, accompanied by a dramatic centralization of the capital markets on which new securities were issued. This was the era of the Sugar and Tobacco Trusts and of the emergence of U.S. Steel. The first of these developments, the emerging industrial concentration, served effectively to inhibit price wars, an important matter because the wars tended to permit a considerable portion of already extracted surplus to be siphoned back to the mass of producer-consumers. This development also signaled a shift, particularly in the United States, toward recognition that the greatest mass of surplus would be extracted from continued productive operations of firms rather than from looting ventures such as had characterized much of nineteenth-century railroad history. The second measure, performed by J. P. Morgan and the other big investment bankers, organized capital markets to support this operation and also simplified market-sharing arrangements and the like through informal regulation of new stock and bond issues.

A revealing development occurred in the United States, namely, the domination by big business of the growing collection of congressional and regulatory "controls" over business activity. Because of the existence of a large and politically relatively organized commercial farming population, the operations of the growing monopoly capital sector did not go unchallenged. A variety of regulative agencies, such as the Interstate Commerce Commission, had been created and in the early twentieth century several of the biggest monopolies were broken up by political and judicial action. These events served as an important educational experience for the monopoly capitalists. They discovered that despite the voting power of the farm bloc, Congress in particular and the government in general remained firmly in their hands. Even in those days it took a fair amount of money to get elected, and monopoly capital had overwhelmingly the largest pool of this in the land. So it was possible to establish an extremely effective political regime. Elected officials were "responsive" to mass voting pressure by passing legislation that appeared to be in the interests of such groups; but as a result of the powers of appointment, appropriation of funds and administrative powers, vested in officials whose future depended on the favor of monopoly capital, this legislation could be kept as only a little more than window dressing. The regulative agencies, for example, often turned out to be a blessing in disguise, permitting more overt collusion and price fixing than would otherwise have been feasible.

And so it was with trust-busting as well. An element of competition among firms, it was found, made for more efficient operation and consequently for more effective surplus extraction. At least that was true so long as prices could be kept quite close to their monopoly levels. But, of course, that was just what the new form of competition was designed to do. And as technology produced new products, monopoly capital could move in, confidently expecting that the combination of the already existing trade arrangements and the regulation of the capital market would prevent competition for shares of new markets from getting out of hand. Consequently, movements in the direction of increasing productivity, instead of being a threat to the system, were merely an effective way to increase the surplus. Thus was born the system of oligopoly in industry that was to dominate the century's surplus-extraction process.

Naturally, the precise chronology of these developments varied from country to country. In Germany concentration was high from the first and investment banks were deeply involved in industrial regulation well before the turn of the century. In Britain these developments came more slowly, the crucial change probably occurring as a result of economic controls and the resulting cooperation among industrialists brought on by the First World War. But long before World War II this basic structure was established in all the major capitalist countries.

Adaptation to the problem posed by the domestic business cycle is a familiar story, and it too has been universalized. The first step involved getting a better understanding of the functioning of money, and particularly of bank credit, in the economy and then developing instruments for effective control. The Keynesian revolution introduced a new set of fiscal instruments and a new theory of control. It took a while to convince many capitalists that this kind of economic intervention should be permitted—indeed it took some of them quite a while even to understand the basic argument—but they finally came around, and monetary-fiscal regulation of the level of economic activity is a standard feature of contemporary monopoly capitalism.

Adaptation to the growing economic interdependence among capitalist nations has been later in coming and far less substantial. The international monetary organizations have no effective power to follow the control procedures used domestically, but liberal theory indicates that that is approximately what is needed. Coordination of monetary policies in the general interest has also been conspicuous by its absence at key crisis points in recent years.

The Pax Americana certainly eliminated significant and overt imperialist rivalries among the great capitalist nations during the postwar period. Of course it was not really much of a pax, given that over fifty wars occurred during the period.[2] But the overwhelming military dominance of the United States within the imperialist camp, together with the felt threat from the Soviet Union, combined to suppress the sort of behavior that dominated the scene before World War I. What this suggests is that an adaptation has not been made in this area, only that circumstances for a while created relative stability. These circumstances are easing in the seventies, and so a test as to whether a significant adaptation of the system has been made is probably in the cards before too long.[3]

Finally, there is the adaptation to the great domestic threat posed by the rise of an organized working class. One factor made life a bit easier for monopoly capital, namely, the continued and substantial upward trend in labor productivity. This allowed resources to be devoted to modest increases in real wages and to the buying off of potential militant leaders within the labor movement. For example, in the United States the building trades have tended to exert far more than proportionate influence both within the labor movement and on the political scene. Their conservative and politically nonmilitant orientation has made them most effective in dealing with the ruling class, since in effect they help to stabilize the basic institutions of monopoly capitalism.[4]

Other adaptations were made in this central area of conflict, an area that did pose a genuine survival threat to the capitalist system.

Union influence within the business sector was accepted, and workers in big business factories were granted a relatively high degree of economic security, medicines well-known for their antimilitancy properties. An elaborate system of division of status among workers was instituted within business and government with a similar aim in view. And the development of military and police technology and training to control domestic disorder was also instituted. Many of these adaptations came later in the United States than elsewhere, but, once again, they are today universal features of the monopoly capitalist scene.

The Dynamics of Lurching

One of the stocks-in-trade of bourgeois economics is the fragmentation of research. By dividing big issues into many little ones, a false sense of manageability is conveyed to the student. Furthermore, some of the most powerful interactive forces in modern society are completely ignored by this approach. That makes for a very comfortable ideology but for a gross distortion of the truth.

Nowhere is this strategy more effective, or more pernicious, than in the liberal attempts to analyze twentieth-century capitalism. The problems of wars, revolutions, even major depressions, all seem to be treated as if they were exogenous; indeed, one can find liberals even heaping praise on the system of advanced capitalism for its ability to withstand the "shocks" that "circumstances" have forced it to endure since the turn of the century.

Nothing, of course, could be further from the truth. Wars, revolutions, depressions, and all the other forms of crisis are endogenous to monopoly capitalism, products of a class society undergoing rapid change that its social science is incapable of understanding. And so the giant lurches from one great catastrophe to another as a fundamental part of its nature. Figure 1 charts key aspects of this lurching movement, which is described in more detail in the following text.

The stabilization of domestic capitalist rivalries via monopolization had proceeded quite far by 1914. But other elements of the crisis remained, in particular the ruinous rivalry between national coalitions of capitalists. Indeed, domestic monopolization may even have intensified this particular contradiction between the class and regional interests of monopoly capital. At any rate, such rivalries, abetted by attempts to use nationalism to defuse the increasingly militant working class, were the dominant causes of the First World War, with its tens of millions of dead, one of the most impersonal of the humanly directed slaughters known to history.

A caused event itself has consequences. World War I is an endogenous part of the world monopoly capitalist system, both caused by it and setting in train forces that were strongly to condition its fur-

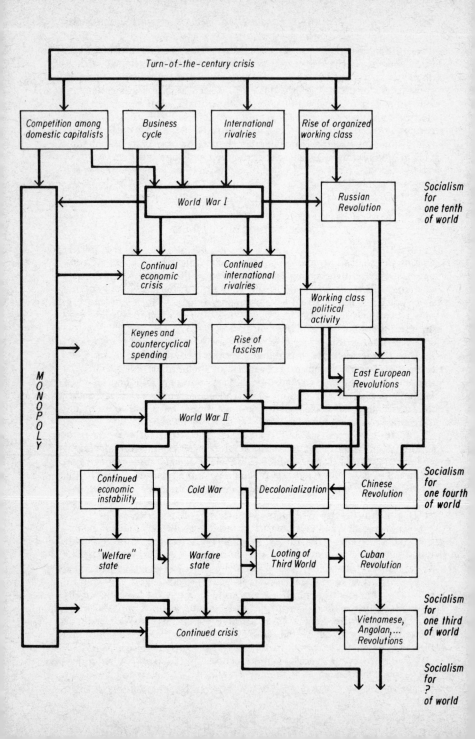

ther development. Perhaps most important of these was the direct political threat represented by war weariness, the collapse of the German
government and the establishment of the world's first socialist government in the Soviet Union. The early stages of the overthrow of capitalism as a whole seemed to have been reached for a few brief months in
1918–19, with revolutionary governments being set up in parts of
Germany, for example. Though the crisis was survived it left a strong
mark. Social legislation and repression were used in varying combinations in the different countries, depending on their internal situation.
For example, the former was predominant in Britain, the latter in the
United States, and Germany used both, combining a social democratic
government with military suppression of revolutionaries. Efforts were
made to overthrow the Soviet government too, but these were beaten
back, and domestic fears prevented any large-scale direct intervention
beyond a few events such as the American occupation of eastern
Siberia, the British occupation of Murmansk, and a French-Polish military advance toward Moscow. As a consequence, some unknown but
considerable portion of the social-security benefits obtained by workers
under monopoly capitalism can be laid at the door of the continuing
specter of a functioning socialist society where the means of production are no longer under private control.

Serious as it was, the Soviet threat was not powerful enough to
unite the national coalitions of capitalism, and international instability
continued to be a major weakness of the system. Attempts to loot the
defeated German coalition by means of "reparations" payments symbolize this continuing conflict, as do the nationalistic trade and monetary policies of the thirties. But the disruptions of the war had major
domestic impact as well. Unemployment was a serious problem in
these economies in the twenties. In some, such as Britain, it was a
desperate problem, and the capitalists could agree on no measures
that would resolve the problem for fear of eroding their own individual
shares of the loot. The trend toward monopoly in an uncertain environment seemed to produce some stagnationist tendencies. The brittle-
ness of a society based on this combination of monopoly capitalist features was revealed in the Great Depression. The resort to fascism
provided a means for the German capitalist coalition to return to its
"rightful" place at the international surplus trough, while other governments struggled against total collapse of their domestic economies.
No relief via acceptable domestic policies seemed to be in sight; for
example, in the later thirties the American economy was heading back
toward the desperate and protorevolutionary situation of 1931 to
1933.[5] Not surprisingly, the technique of international scapegoating
suggested itself as a solution, and belligerency in foreign policy began
to increase markedly. The United States "abandoned" neutrality, sent
destroyers to Britain, and decreed a boycott on Japanese import of
vital American scrap iron. After all, this solution—escape from revolution through war and preparations for war—had worked before, both in
1933 in Germany and in 1914 all over Europe.

And in a sense it worked again. Unemployment was ended for a time, as was domestic class conflict. However, world war is a rather myopic policy from the point of view of the survival of monopoly capitalism. The end of the war found the European capitalist economies in a state of deep crisis. American economists and politicians were in general agreement that, shortly after the war's end, the American economy would probably sink back into a depression similar to that of the thirties. Colonialism was plainly on its last legs, promising a fundamental readjustment in the relations between imperial and dependent states. Domestic communism posed a serious challenge to the French and Italian governments. And not only was the Soviet Union a victorious power in the war, having assisted in the new defeat of the German and Japanese coalitions of capitalists, but socialism was again on the march, heading for power in a number of East European and East Asian countries.

Once again the solution to one crisis sent the monopoly capitalist system lurching off into another. However, the solution tried this time was more effective, at least for a while, than had been some previous ones. For one thing, the Keynesian lesson had been learned, and by and large monopoly capitalists had come to accept the idea that a program of economic stabilization could be designed so as to increase the surplus accruing to each major capitalist group. Second, the military and political dominance acquired by the United States served, for a while at least, to eliminate one of the major weaknesses of the system, namely, imperial rivalries. In fact the United States, acting partly to support domestic capitalists by providing them with orders and partly to meet the threat of the Soviet Union, offered substantial economic aid to help start up the European economies once again.

Another major response was the Cold War. This too had several dimensions. First, the Soviet Union, of course, by its very existence posed a threat to the capitalist system, though the threat was much weakened by war losses and by Stalinism in the late forties. Second, there was the fear that communism could not be contained before it swept across the decolonializing Third World, and even perhaps southern Europe as well, where a civil war in Greece was already underway. Third, there was now ample evidence that government military spending could be a very profitable business, while war tended to blunt domestic class conflict. The Cold War was a more efficient version of a hot war, permitting the suppression of dissent in the interests of "national survival" ("clear and present danger" was the phrase much used by liberal Supreme Court justices), while at the same time permitting the economy to operate without inconvenient economic controls.

This system worked reasonably well, from the point of view of the monopoly capitalists, for a decade or two. It even provided one or two bonuses, which had perhaps not been fully appreciated at the start. The large-scale military spending made massive resources available to control the process of decolonialization in such a way as to be minimally disturbing to established capitalist relations in the Third World; at the same time the spread of "communism" was checked for a while after it had reached out to cover not much more than a fourth of the world's population. Indeed, the opportunities for extracting surplus from the newly emerging Third World may even have been increased, as the national bourgeoisie in these countries could be used as a screen for monopoly capitalist operations. The Pax Americana was a bloody and frightening time, full of war and misery for hundreds of millions of the world's citizens, but from the point of view of monopoly capitalism it was a stable and prosperous period.

The Current Crisis

To paraphrase one of those Hegelian triads, with monopoly capitalism success breeds failure. The Pax Americana was in a certain sense a time of stability and prosperity for monopoly capitalists, but by now at least the stability side of it has come to a close. Europe and Japan have been playing successful economic catch-up to the point at which they can once again begin to challenge the American capitalists. This affects stability in the same two basic ways. First, it makes international financial control difficult, as complex negotiations affecting different coalitions have to be worked out in a rapidly changing situation. And, second, it makes the renewal of imperialist rivalries feasible as Japan, Germany, and other groups accelerate their foreign-investment programs.

Success has bred failure in other areas too, notably in the challenge that some developing countries are able to offer monopoly capital. The oil crisis needs no comment, and however the oil-producer consortium works out it suggests fundamental problems of adaptation among competing regional coalitions of capital.

Yet another area in which success has tended to breed failure is that of the new government-business alliance. This alliance offered opportunities to buy off labor and to socialize some of the costs of doing business, while at the same time stabilizing the economy. However, it has developed instead into an additional arena for combat among competing capitals. A consequence of this has been the development of inflation, whose rate has actually been on the increase for over a decade. And in political-economic terms, inflation is far more difficult than depression to cure, because cutbacks are difficult to allocate among the competing domestic capital coalitions. Those highly touted seven or eight years of steadily rising real GNP in the sixties were preparing the way for the instability and uncertainty of the seventies.

Another frightening underlying aspect of the Pax Americana has been the steady militarization of the world. Not only is there more than one gun per capita in private ownership in the United States, but American and other nations' arms sales, direct and indirect, have spread weapons of destruction and the capacity to produce them in truly massive quantities all around the world. The result has been that the cost of war has gone up for all sides, as have the expected levels of destruction that a war will achieve. The burgeoning crews of military technicians, of citizens trained in the arts of war, of the military in high political positions, all suggest that war is much more on the minds of the world's leaders these days.

The waste, the irrationality, the insanity of this, the most stable era of monopoly capitalism, cannot but affect deeply the lives and attitudes of everyone who willy-nilly is a part of it. The malaise of life in urban and suburban America needs no retelling here, but surely it reflects the generalization of Marxian production-process alienation to every corner of life in monopoly capitalist society. The revolt of youth in the sixties was but one—though at least a rational one—of the manifestations of this deep sickness in society. Out of the disenchantment with traditional liberal ideals that spread so widely during the sixties will perhaps come both the understanding and the commitment to toss this intolerable social system into the historical ashcan.

CHAPTER 5

Exploitation Under Monopoly Capitalism

THE AVAILABLE EVIDENCE on human history shows that when a surplus begins to be produced in a society, attempts will be made to extract that surplus from the producers for the use of others. It seems natural enough to call this exploitation. The success of the attempts tends to vary over time and place and the method of extraction used. There does appear to be a very broad and powerful tendency for the rate of extraction to increase more rapidly than the productivity of labor; that is, as societies become more productive, the direct producers tend to benefit relatively less from the improvement than do those to whom the extracted surplus is allocated.

It is important to note that we are commenting on a widely observed historical tendency, not on some rigidly deterministic theory. Throughout history nonexploitative microsocieties have coexisted with exploitative ones. Some of these nonexploitative societies have been too poor for a significant surplus to exist. In the early nineteenth century the western Shoshone Indians searched in nuclear-family groups for acorns and bugs as their basic diet. Occasionally they came together to hunt rabbit or fish. At such a time some surplus over essential consumption no doubt was often created. Just as obviously, this surplus was typically retained by the quasi-community of producers. Relatively low productivity or technical difficulties in surplus extraction seem to have been the main reasons for the survival of such groups; of course, the same factors tend also to account for their relatively small numbers.

Among societies where surplus extraction was practiced as a central feature of social life, the particular extraction process also varied widely over time and place. This was noted in chapter 3, but it becomes relevant again as we look at this aspect of monopoly capitalism. There has been an important change in the structure of exploitation with the shift from capitalism's nineteenth- to its twentieth-century versions.

The industrial revolution wrought a great change in the opportunities for surplus extraction. Exploitation in peasant societies was based essentially on the control over land exercised by a ruling class, but this process was affected by the relatively close ties that tended to exist between the producer himself and the land, stemming both from cultural considerations and the uniqueness of individual plots of land. The industrial revolution brings produced means of production to the fore as the control instrument for surplus extraction. But these capital goods are not unique in the same sense; items of machinery are often mutually interchangeable, and most any factory is reproducible. This made it possible to wholly separate the producer from any long-run connection with the basic means of production he was to use.

Out of this grew the generalized system of wage labor, and with it the so-called law of value, a concept that gives some quantitative expression to this highly impersonal system of surplus extraction. For a given state of the forces of production the new market system tended to generate, in the long run, prices that reflected the labor content of the goods produced. This is a simple consequence of the long-run reproducibility of goods, including capital goods and labor, and continues to hold so long as there is no major resource exhaustion. Ruling-class control of the means of production permitted that class to keep the price, the wage, the exchange value of labor at a level that, roughly, was minimally consistent with productivity and the replacement of aging workers. The surplus accrued to the capitalists by letting the market compel the worker to work for much longer each day than was necessary to recreate the value of his labor. The actual rate of surplus extraction might vary from industry to industry, but the process of extraction was governed by these same impersonal factors, operating under the primary influence of the capitalist class's monopoly control of the means of production. The worker's choice was simple: Work for very low, monopolist-induced wages or starve.

Of course, this was not a wholly novel process of surplus extraction. Factories and shops employing labor for wages existed in many earlier societies. However, in these earlier instances the process was embedded in a very different kind of society, and this affected many aspects of its operation. For example, there was often a paternalistic element to these earlier forms, though this may have meant in human terms no more than that there were social processes oriented toward establishing feelings of loyalty in the worker and that these feelings were one of the devices by which surplus was extracted. At any rate, the newly generalized system of wage labor essentially eliminated such master/servant ties. And this in turn was one of the factors generating alienation and fractured personalities in the new society, as people were impelled by the new environment to treat one another as mutually interchangeable and exchangeable objects.[1]

There are, of course, many other consequences of this transformation of the central process of resource extraction in modern society. But enough has been said to suggest the fundamental aspects of the transformation. This process continues to operate in contemporary monopoly capitalism, and remains a fundamental source of surplus for the ruling class. However, monopoly capitalism has brought new processes of surplus extraction to the fore and has brought about yet another generalization of the process of surplus extraction. It is to these new processes that we must now turn.

Surplus and Its Growth

Celebrations of capitalism always place rates of growth of output as the centerpiece of accomplishment. Since the turn of the century in the United States, this story will go, output per capita has increased almost fourfold. Undeniably that is a great accomplishment, signifying a genuine increase in the level of the forces of production and offering the technical possibility for tremendous further accomplishments that are more directly relevant for human beings.

Unfortunately, the measure of output used by economists is hard to relate to the ways in which human lives are changed by the output increase. That is precisely because it is not really a measure of achievement but of promise, of what might be done. The distinction was brought out most clearly in a little calculation made many years ago by Paul Baran that relates to the changes wrought in the American economy during World War II. In 1944, for the first time since World War I, there was approximate full employment. However, of the 66 million members of the officially designated labor force, some 11 million were in the armed forces and thus not engaged in productive labor. And about half of the officially designated output produced by the remaining four-fifths of the labor force was military goods. This massive mobilization and reordering of the economy from its gross malfunctioning of the depressed thirties, you would think, must have called for a tremendous material sacrifice by the population. But there are strong indications that reality was very different, and that in fact the general standard of living of the mass of the population actually improved. The implication is clear: The American population could be brought to a peak standard of living by using only half the "output" of the economy. Output per capita has at least doubled in the thirty years since Baran's calculation. Also, in those days not much more than a third of working-age women were officially in the labor force; nowadays the figure is more like one-half. When one adds these facts in, and makes some allowance for the need to maintain and perhaps increase the stock of factories and durable goods in the economy, it would be a conservative estimate that our current population could be maintained at a decent, "American" standard of living with the use of well under a half of our "output." [2]

That is quite an astonishing figure. We are deluged these days with complaints and comments by politicians and pundits to the effect that "there is no free lunch," that there is no way we can provide for the environment or for the poor. We must live, even our "liberal" politicians tell us, for years with our 25 million officially poor, people who by no stretch of the imagination have a decent standard of living. We must also live with inadequate medical service, with poor schools, with commodity shortages, even with inadequate and overcrowded jails. What are they doing with that half to two-thirds of our potential output?

It's a good question. You will not find the answer in any official statistical table, because such figures are not considered respectable by our eminently respectable economists. As a consequence, we can't put numbers to the various amounts of misuse of our economic potential. However, we can at least make a list of the things that happen to prevent the tremendous promise heralded in the "output" figures from becoming achievements. As we go through this list, it might be well to remember not only that there is this tremendous promise in the United States, still the world's richest country in its economic potential, but that there is really no social accounting for the goods that represent that potential. We are reduced to qualitative and speculative comments precisely because this great secret is being kept from the American people by a massive obfuscatory flood of irrelevant statistics and commentary by politicians and economists alike.

The question we have just referred to is, How can we resolve the contradiction of scarcity in the midst of plenty? On one side we have tremendous productive capacity, holding out the promise of a society of abundance; on the other side we have tremendous unmet needs, needs that seem to be getting more intense even as the capacity to satisfy them grows. On one side we have the economists' statistics that show "consumption per capita" rising more than three-fold over the era of monopoly capitalism in the United States; on the other we find tens of millions in poverty, children hungry and their illnesses untended, and all around us signs of a deteriorating environment. Of coure we want to know why such things happen; but first there is the simple question of fact: What is happening, where is the promise gone? Here are some clues to the mystery.

1. Quite a bit of that potential output is simply unproduced. In an average peacetime year between 5 and 10 percent of the labor force is out of work. Average use of our stock of capital is at a substantially lower level. Because of the inherent instability of the capitalist system and of the opportunity for big profits during the brief periods of frenetic prosperity, the United States uses its stock of productive capital at an average rate of only three-fourths or so of potential. Given the expansion of the last three decades, this means that the currently *unused* plant and workers in a typical year might have produced close to half of the massive surplus generated in 1944.

2. There is a tremendous amount of wasteful production. The magnitude is suggested by a study that found that the cost of model changes in producing Ford cars over a four-year period came to over

a quarter of the purchase price of the car.[3] Basically, these costs relate to getting people to buy more cars sooner, not to improvements in the functioning of the car as a means of transportation; they represent waste. The drug industry is another one that is notorious for generating wastes of the above kind. In addition, it turns out that only about six cents on every drug sales' dollar goes to research, while twenty-four cents goes to advertising.[4] Highway accidents often produce fifty thousand annual deaths and a grand total of a million annual casualties. These damaged goods and people, of course, have to be repaired, and the repair itself is an expensive process. The above items are a small sampling of things that divert the promise of our productive capacity to waste, for they are largely avoidable and represent burdens on our capacity brought about by mismanagement.

3. In addition to wasteful production, one can distinguish wasteful occupations. That is, many jobs that contribute "value added" to the official national product use talented human beings in essentially unproductive ways. Advertising has already been mentioned. Nowadays most nonlawyers would recognize a large fraction of that crew as wastefully employed protecting clients against other lawyers' actions and generally working massively at paper combat for pay. The proliferation of bureaucracies everywhere in the towering office buildings of downtown America are eloquent testimony to the wastefulness of American business and government. Even the marketplace, supposedly a paragon of efficiency, turns out to be tremendously wasteful, given the costs of generating and organizing information in its anarchic mode of operation and in peddling the goods.[5] Literally millions of jobs are filled in these activities and, more than likely, most of them are unnecessary.

4. There is now a growing awareness of the tremendous amount of loss-producing economic activity in the United States. Urban sprawl reduces the amount of available agricultural land, and the cost of producing farm products rises as a result. The transport of goods and people and the production of goods by unsafe methods causes casualties that must be repaired. Among the affluent and well-insured the doctors prey, providing unneeded and even pernicious services, for example by performing truly massive amounts of unneeded surgery. And, of course, pollution by some factories raises the cost of production in others, and generates debilitating respiratory and other ailments in surrounding populations. Under the incentives of modern business it has often been less costly to spill oil than to take simple and even cheap preventive measures.

5. Perhaps a separate category should be allotted to the three major illegal activities, drugs, prostitution, and gambling. Losses of that promised output resulting from unproductive and degrading aspects of these activities are unmeasured but clearly very substantial.

6. It is no secret that our income distribution is highly skewed, far more so than is necessary as a material incentive for workers to generate the potential output in the economy. There are hundreds of individuals in our country who report over a million dollars of annual

income, and probably hundreds more who have it but don't report it. Over a quarter of personal income goes to less than 10 percent of the families. And perhaps three fourths or more of the privately held securities are owned by less than 2 percent of the families.[6] The diversion of our productive promise to servicing the fashionable "needs" of the wealthy, through both the private market and government favor, represents a substantial fraction of potential output.

7.   There is also a major bit of information as to where the surplus did *not* go. In the last dozen years or so it did not go to the working class, since their real wages have remained stagnant.[7]

8.   But there is a substantial fiction even in the statistics that show a rise in the real incomes of families in the broad mid-range of the income distribution. One telling figure is the virtual stagnation in the average amount of time worked over the last thirty years. On the face of it, this suggests that rising incomes do not permit the population to take more leisure, but that they must continue to work about as much as before simply to stay even. Much of what we think of as income increase is really simply compensating us for losses that our new environment has created. Most obvious perhaps is the expenditure for transportation. Expensive cars must be increasingly used as urban public transport deteriorates. Movement to places farther away from the job is made necessary by the decay of formerly attractive neighborhoods in town. Because of this a car becomes a necessity for security, for shopping, for recreation, where in the past these needs could be conveniently met locally. Despite all those expenditures on fancy appliances, studies indicate that housewives do not spend less time on housework, suggesting that once again the gains from the goods are eaten up by compensating expenditures.[8] Some measured increases in "real" income are fictitious, as when rising wages increase the "value" of essentially the same services provided earlier for less. The low quality and durability of many goods, substantially less than in the past or at least quite low compared with their "promise," mean that another fraction of that increased income goes to maintenance and replacement. Perhaps the most striking and important single example of this phenomenon of compensating expenditure lies in the field of health care, where expenditures are several times greater than in past years when the measured health status of this mid-range of the population was comparable to its present state.

9.   Finally, there is the category of alienation expenditures. This would be the hardest to quantify, but is probably also the most pervasive, tainting almost the full range of human activity in contemporary America. Job absenteeism, reduced performance by the millions of citizens on Valium and booze, these are measurable consequences of the alienating environment generated in our society. Beyond lie less clear-cut but still obvious enough phenomena. For example, the increasing failure of our schools is clearly related to their misuse as institutions for advanced baby-sitting and disciplining of youth. The "successful" product of these schools is basically certified as someone who is willing to spend endless time doing meaningless things on order. Work

has been moving in the same direction, with specialization and routinization depriving most participants of any sense of creativity or personal responsibility for the product.[9] Surrounding everyone—from the first look at television, the first day in school, the first day on the job —is the brainwashing, the incitement to pitch in and make America great by first working harder and then consuming harder. The costs of creating a whole society that is thus forced to accelerate with its brakes on is the last great sink down which our productive promise flows. One might seriously question whether the last generation or two of life under monopoly capital has produced any real increase in income at all.

Surplus and Insecurity

Defenders of monopoly capitalism have another line of defense set up behind the breached barriers of the growth ethos. This is the argument that the system has provided a tremendous increase in the security of its citizens. Social security, unemployment compensation, medical insurance: In these and other ways monopoly capitalism, it is claimed, has reduced life's risks, thus making our existence a good deal more pleasant than levels of real income might indicate.

Like other such arguments, this one is based on a half truth that grossly distorts the reality of the situation. The above programs do exist and provide coverage for a substantial fraction of the population, though the fraction varies quite a bit from one program to another. In addition, improvements in the state of the art of medicine have led to a very substantial reduction in the death rates of young children, certainly a major improvement that can be laid in part at the door of the political economic system. In this case it did in fact deliver the goods, even if still on a selective basis, providing, for example, much higher death rates for poor and minority members of the population than for rich and white members.[10]

But the half of the truth submerged in this account is not so difficult to dredge up. Capitalism is still a highly volatile economic system. Unemployment rates are both high and variable, by no means everyone is covered by unemployment insurance, and incomes tend to drop substantially when a member of the family is laid off.[11] Inflation too is high and variable and eats at the other end of the income, so to speak, producing at times large drops in the ability of families to purchase their accustomed mix of goods. The variability of commodity prices, which basically reflects the intensifying conflict between monopoly capital and the Third World, is another large source of income insecurity, affecting both the prices of goods and many jobs.

Much of the insecurity of modern life is a direct product of monopoly capitalism, all right, but is less directly connected to economic processes of the marketplace. The virtual explosion of crime, by no means restricted to cities or to the United States, is directly related to the political economy of monopoly capitalism: It breeds on poverty and alienation and hopelessness. The ruin of neighborhoods, even of whole cities, is a serious risk under mature monopoly capitalism, one for which no insurance can be bought and that government policy seems strangely incapable of abating. The lower risks to traditional causes of infant deaths are at least partly compensated for by the higher risks to the general population from cancer and respiratory ailments, whose increased incidence now seems to be quite closely tied to the deteriorating environment created by monopoly capital's great organizations. And the risk of death from war has never before in history been so high, including the serious prospect of the destruction of all human society. The age of monopoly capital is *not* an age of security, not even for Americans.

Surrounding and permeating human behavior under all these risks, giving that behavior its bizarre hues of violence and of madness, is the depersonalization of human relations under monopoly capital. We become not so much predators as objects to one another. This in turn deprives the individual of that elemental security that derives from being a member of a community, of a group of people who have some commitment for mutual aid and support of one another. The deterioration of security in this dimension may be the most fundamental risk of all. Out of it grows the tolerance for misery in others that conditions the growing risks in contemporary society. And it in turn grows out of the system of monopoly capital itself, which forces depersonalization on participants as a condition for survival within the system.

Exploitation and Monopoly Capital

This chapter began with a brief account of the classical Marxian processes of exploitation. But most of the chapter was devoted to appraising the allocation of the surplus in contemporary monopoly capitalism, in particular the various diversions from its effective use and the ways in which risks relevant to the citizen have changed. Does this mean that the subject has been changed, that we are no longer speaking of exploitation?

I think not. Exploitation in the traditional sense, the extraction of surplus from the producers by a ruling class, continues to be a central feature of monopoly capitalism. But the classical Marxist version, based on labor reproduction costs under a labor theory of value, is no longer capable of telling the whole story of exploitation under monopoly capitalism. A wholly satisfactory alternative explanation is not yet forthcoming. What we are trying to do here is to suggest some elements that must be included in such an account.

The basic thrust of the argument does not imply a rejection of Marx's theory of exploitation.[12] Our society can eliminate material deprivation as a cause of human misery. Indeed, the Baranian calculation suggests that, given time to make appropriate adaptations in our stock of capital and our institutions, we have the *present* capacity to achieve this. The promise remains unrealized, and appears to many to be unrealizable, because of the fundamental flaw in our society, which produces all those diversions of resources and increased risks of ruin. That flaw is the class nature of the society, which generates the endless series of conflicts, negotiations, truces, chicanery, and renewed conflicts that destroys the society's potential. Only by creating a new society can we achieve the promise and eliminate material deprivation as a cause of human misery. This is the only real resolution of the contradiction of scarcity in the midst of plenty.

But, clearly, contemporary exploitation differs from its predecessors. It is the incredible waste of resources that strikes the eye here, and the distortion of human behavior, rather than any quantum of material deprivation, though the latter continues to exist for many millions. True, the ruling class continues to expropriate a very substantial portion of the mammoth surplus for its own uses; but in this weird society the ruling class is itself to some extent a victim as well as an exploiter. The lives of its members too are distorted by the society's inhumane values and the behavior that is required to sustain the system. Alienation deprives many goods of much of their value for any consumer. But this in itself offers hope, as it has now become possible to recruit substantial numbers of the young from the ruling class itself to the goals of creating a new society, one in whose success they too have a direct stake.

CHAPTER 6

Government and Monopoly Capital

IN THE DAYS of classical capitalism, the functions of state and economy were fairly sharply separated. In the more advantaged capitalist countries, such as nineteenth-century Britain and the United States, one might reasonably say that the primary function of the state was to protect the legal-institutional system of private property relations. These relations guaranteed bourgeois control over the means of production and thus determined the nature of exploitation of the direct producers, which in turn was a product of ruling-class control over the produced means of production. This monopoly, of course, is the key to understanding the operation of what liberals call the free market. The state is required because ownership is quite an intangible thing, defined at law in terms of the phrasing of certain documents and protected by the threat of use of the coercive instruments under the state's control. The state could thus remain in the background, available for very tangible use whenever the "rights" and power of the bourgeoisie were called into question, but otherwise very much less active than the "private" sector of the economy, where nearly all production and exchange occurred.

Things are very different nowadays, at least on the surface. No one could accuse the governments of monopoly capitalism either of being in the background or of being sharply separated from the economy. In trying to understand these changes, it is vital not to take the static, snapshot approach of simply comparing the early capitalist state with the monopoly capitalist state. Though each of these regimes has its own special properties, understanding is only gained by looking at the process by which the properties were acquired. Here as elsewhere change is to be understood as the resultant of struggle, often of violent struggle, among a variety of contenders for power. The struggle shapes the outcomes and deprives those outcomes of their rationality,

however "rational" may have been the strategies of individual contend-
ing factions. And, of course, the struggle continues, shaped both by its
past history and by its present environment. In our brief look at the
process of development of the monopoly capital state we will find one
central, invariant theme that ties the early and the late capitalist state
together: The primary function of the state continues to be the protection of the regime of private property relations.

The Rise of Big Government

At the turn of the century federal, state, and local government expenditures in the United States were $1.7 billion, or about 10 percent of
gross national product. Nowadays the government expenditures figure
is around $600 billion, or about 40 percent of the current value of
gross national product. That is quite a change, obviously one that has
had a dramatic influence on the behavior of both the state and the
economy. How did it come about?

At the top of any list of explanations of the rise of big government in the twentieth century is war. The United States illustrates
this process very well. Military expenditures expand at a good pace
during the prewar years of increasing international tension (a 50 percent increase in 1902 to 1913 and a trebling in 1934 to 1940), then
explode during the war (about twentyfold for World War I and tenfold
for World War II). During the course of this explosion, business grows
accustomed to military contracts and finds ways for mutually benefi-
cial collaboration among businessmen who in the past had been at
loggerheads. Consequently, the new peacetime levels of expenditures
are far higher than those of even the tense prewar period (1922 military expenditures were more than three times those of 1913, and 1948
more than six times those of 1940).[1] As we saw in chapter 3, the wars
themselves are no accidents of history but are natural products of the
functioning of the capitalist system. So indeed is this internal manifestation of permanently rising military expenditures.

The experience of war teaches businessmen not to hate war, or
especially to love it, but to appreciate the opportunities inherent in
close interaction with government. But this means a different kind of
government-business relationship. Government is no longer just protecting property rights and mediating conflicts among business groups.
It is now collecting surplus through taxation, a form of coercive ex-
propriation, and then allotting the surplus to various capitalists in the
form of military contracts. Of course, some of this has always gone
on. But at these new levels of expenditure the government becomes by
far the largest demander of goods in the economy. Instead of mediating conflict among businessmen, it now becomes something more like
its feudal counterpart, a direct participant in the squabble within the

ruling class for a share of the surplus loot. Clearly, this represents a change of some significance in the relations of production.

The second major change in the role of government is a product of the increasing need for social-harmony expenditures, that is, for expenditures that reduce the risk of overthrow of the capitalist system. The first major increase in this sector of expenditures came in education. Expenditures on education have not only increased rapidly throughout the twentieth century in the United States, but during pre-World War II peacetime years they tended to be at least two or three times greater than national "defense" expenditures. Some of this increase is associated with a genuine need for a higher level of technical training for workers in modern industry. But as the origins of the movement for universal compulsory education indicate, and as the ideological slant of the textbooks—adopted for compulsory use—confirm, the principal motivation in this area was control and discipline of the work force. Early increases are associated with rapid urbanization and the threats posed by the concentration of large numbers of workers in towns. Increasing literacy does not seem to have been the central motivation; for example, in Massachusetts the districts whose factories required little skill tended to opt for compulsory primary education before those employing a large fraction of skilled labor.[2]

A second and later thrust of social-harmony expenditures is of course associated with the word "welfare." It should be noted that programs primarily supplying relief for the destitute have never assumed a large role either absolutely or relatively in government spending. There seems to be some feeling that stabilizing a large idle force of working-class citizens is not the way to go about preventing revolutionary ideas from spreading. Instead, the welfare budgets have increasingly been allotted to a variety of programs aimed, at least minimally, at dealing with specific problems of the poor. There has been much public housing construction, many hospitals have been built, urban services expanded to create jobs, and subsidies provided to assist the purchase of food, health, housing, and other specific services by the poor. Both relatively and absolutely the great expansions of this kind occurred in the United States after World War II, though other capitalist countries began the process somewhat earlier.

By now it is possible to discover from government statistical documents that expenditures on welfare-related items substantially exceed national defense expenditures. Monopoly capitalist governments are now often referred to as "the welfare state" and defended as instruments of mercy in a threatening world. Unfortunately, the facts belie this kind assessment. One only need remember that such expenditures stem from the nature of capitalism, a system in which the profit-oriented ruling class does not assign tens of billions of dollars of the surplus it has extracted from the direct producers as a simple act of mercy. The first impetus for such spending, as noted, comes from the fear of revolution. The ruling class is very small—substantially less than a tenth of the population on any reasonable measure. When things go wrong with the system, as during the Great Depression of the thirties, fear becomes a major factor in ruling-class reactions. In a number of capitalist societies, this reaction expressed itself in the rise of fascism. In the United States and Great Britain, the opportunity was seen to preserve power by less drastic means. And thus the "welfare" program was born. As already noted, fear of the Soviet Union also contributed to monopoly capitalist attitudes.

However, once instituted, such programs tend to take on a life of their own, again conditioned by the nature of the system. They became structured in such a way that the affluent could profit from "welfare" as much as or even more than the supposed "target populations." It has been estimated that the incomes of medical doctors increased by more than 50 percent within a very few years of the introduction of Medicare. Opportunities to profit from all sorts of construction contracts associated with welfare programs have been legion. And even job programs have a strong effect on the system. There are now several hundred thousand officials administering poverty programs.[3] The allegiance of these people to the government that employs them is a distinct plus for the ruling class. Furthermore, they now have a stake in the complex and unwieldy service systems of which they are a part and become an influence persistently working for increase in their size and scope.

What is the upshot of all this? At the present time the total welfare budget amounts to around $150 billion. If that money were simply handed out to the poorest people in the country, it would mean a gift of $3000 per year to every man, woman, and child in the poorest quarter of the population. In other words, poverty could be eliminated completely and permanently with a fraction of this sum, even given the general inefficiencies in the capitalist system. But the welfare system is so permeated with waste, corruption, and inefficiency, with rakeoffs to the nonpoor, that poverty itself remains only lightly touched by the trickle of funds that survives all these diversions. One is reminded of the Syr Darya River in Central Asia; as thriving cities grew along its banks, and with them their needs for water, the river flow steadily diminished and finally disappeared into the sands long before reaching the sea; and then the cities died.

But in a way there is an even more fundamental consequence of this development. Government and economy now become thoroughly mixed in the process of extracting and allocating surplus to the "welfare" program. Instead of simply laying down and enforcing the general rules of operation of the market economy, we now find as much as a fifth of our output funneled through the government to the welfare and defense systems. Government agencies, businesses, and elected officials are locked in continuous struggle on literally hundreds of fronts, as the various interests scramble for their share of the surplus that has been extracted from the direct producers. The outcomes of these little struggles cannot be socially effective, because the participants are not the supposed beneficiaries. They will not be very much concerned with saving on taxes because each individual program is just a drop in the bucket, given the size of government and the ability of the participants to shift the tax burden away from themselves. Powerful processes have thus been set into motion that will tend to increase both the size of the program and, *pari passu*, its level of waste, of failure to fulfill the stated objectives. It is also a very divisive, conflictful process, even for the participants. And, finally, awareness cannot help but grow among the supposed beneficiaries as to what is really happening to the "welfare" program.

There is a third major cause of the change in government's size and role in monopoly capitalism. This stems from the basic contradiction of capitalism, that between the increasing socialization of the forces of production and the private nature of the relations of production. What has been happening increasingly, as twentieth-century science and technology have continued to develop rapidly, is that the decentralized market has become a steadily less effective mediator of economic activity. If there were such a thing as stable markets, private competitive capitalism could at least have been an efficient, if exploitative, economic system. But new technologies have required a larger scale of operation and, increasingly, more complex and direct relations among producers than the arms-length bargaining the classical market processes could permit. And, of course, modern production has increasingly been generating harmful side effects on the environment that cannot be abated by resort to market processes.

Part of monopoly capital's response has been centralization of private economic organizations. In the United States this has meant, in just the last few decades, almost a doubling of the share of output coming from the largest corporations, until, for example, the top 200 manufacturing firms now control almost three-fifths of the assets employed in manufacturing.[4] Similar—in fact, probably even greater— concentration exists in firms involved mainly in financial operations, such as banks and insurance companies. To the outsider's casual glance these firms begin to look increasingly like a single entity.

But they still retain, in both legal fact and actual practice, their separate identities and autonomy. Consequently there has been an increasingly urgent need to provide mediators and arbitrators for the disputes that continually arise among these giant organizations. Government has an obvious role to play here and, through regulatory agencies and in many other ways, has been serving that role to a rapidly increasing extent. However, government, as noted above, can no longer assume the position of "outside" agent in resolving such conflicts. It is also the economy's largest dispenser of largesse to these same corporations. So increasingly government itself has become a battleground in which corporations vie with one another for a larger slice of the pie. And so the basic contradiction is further intensified, waste and corruption continue to increase, and the conflict-resolution process begins to lose much of its legitimacy. Once the battle over Richard Nixon's survival came to the crunch, the decision tended to be transformed into just such a conflict; the ruling class was itself divided over the appropriate role of government in this confusing environment of intensifying late-capitalist struggle.

These three factors then have primarily conditioned the rise of modern monopoly capitalist government: war, the need for social harmony expenditures, and the increasing socialization of production. They have turned government into a Kafkaesque madhouse of inefficiency, in which its typical functions are almost unrecognizable. Does that mean that an entirely new regime has been created? That is the question to which we now turn.

Normal Government and Crisis Government

Most of the time the monopoly capitalist state appears as the arena for innumerable conflicts among a large number of disparate interests. Business, of course, appears very prominently among these interests. But many of the conflicts are between one segment of business and another. Labor appears frequently in the arena, mostly in conflict with business, but organized labor at times can be found in collaboration with segments of business. Other interests too appear on the scene, from environmentalists to genuine representatives of the impoverished and the minorities. It is perhaps not surprising that some observers have called this process of conflict "democracy."

But that *is* a serious error. In the first place, such a claim fails to take into account the fact that all this conflict takes place within a context of acceptance of the basic operation of the capitalist system. Private property relations continue to be the basis for surplus extraction and allocation. Government actions serve to define new areas of surplus extraction, it is true, but as already noted, the process takes place in a context of continued dominance by the same ruling class, and private property remains the one principal source of personal wealth.

Related to this is the fact that the conflicts themselves are individually small in scale, mere skirmishes over marginal restructurings of the surplus process or, more likely, marginal deviations from proportional divvying up of the growth dividends. From time to time these conflicts are resolved against the interests of some segment of business. But what is lost are mostly skirmishes, only rarely are they battles, and never campaigns. When the results of several years of conflict are tallied up, the ruling class, operating through its control of all the major institutions of production and distribution, finds itself still firmly in the saddle. Consequently, a short-run viewpoint tends to produce two kinds of mistaken feelings: that many groups representing nonruling class interests are very powerful in monopoly capitalism; and that the system is relatively stable. Generating these mistaken feelings is the principal task of capitalism's apologists; an aura of scientific legitimacy is lent their work precisely by their emphasis on amassing the trivia of the daily operation of the system. All those numbers and little facts look very impressive and screen the casual viewer from the real basis of operation of the system.

But the fundamental instability of the system, whose surface manifestation is that dynamics of lurching we have already discussed, on occasion reveals itself in an internal crisis. At such times one gets a glimpse of the true structure of power in the system. The crisis of New York City has been especially revealing in this regard. For long New York has been regarded as a bastion of humane liberalism, with a government fully under the control of the people, supported by powerful organized labor, welfare groups, and other natural opponents of big capital. New York's wage structure and welfare system were apparent reflections of that power, and the high levels of pay and services, relative to the rest of the country, had grown up as the result of a long series of struggles of the kind just described.

But then the "fiscal crunch" came, the city was no longer able to find rich investors to buy its bonds, and a full-scale crisis in this sector of monopoly capitalist government came. What was the upshot? Well, first things first; and that means finding out who's in charge. Just so there would be no more mistakes on that score, this bastion of people's power was immediately taken over by a committee of bankers. Without a shot being fired the power to make budget decisions was taken away from the elected officials and placed in the hands of . . . a committee of the bourgeoisie! Once that was settled, the rest came as might be expected. Wage freezes, massive layoffs, service and welfare cutbacks. This was a minicrisis, affecting only a part of the system. But it represents a straightforward revelation of the true structure of power under contemporary monopoly capitalism.

Is it true then that somewhere in the background there actually is some shadowy supercommittee of the bourgeoisie running the show? The answer to that is no. The functions of the "committee" are handled in normal times by the structure of the system, by that basic set of private property relations to which all participants in the system must pay their obeisance. But when crisis comes one sees clearly enough the secondary and basically legitimizing role that the trappings of democracy play in the United States and elsewhere under monopoly capitalism. In their public statements capitalists in the United States usually pay homage to the democratic process. But that allegiance is only skin deep. For example, during the Watergate period, a crisis period but not one in which the survival of monopoly capitalism was at issue, one found businessmen saying things like this to one another: "This recession will bring about the healthy respect for economic values that the Depression did"; "I can't believe that social responsibility was ever invented by a businessman; it must have been made up by a sociologist." "It is up to each of us, not to some prostitute of a Congressman pandering to get reelected, to decide what should be done." "Maybe we should take the franchise away from government employees so the system can be preserved." [5] The distance between these actually expressed thoughts and fascism is not so very long.

So we find confirmation of the assertion, made at the beginning of this chapter, that under monopoly capitalism the primary function of the state continues to be the preservation of the system of private property relations. The state grows larger, the arena of struggle grows wider, inefficiency and corruption flourish more luxuriantly, as the history of the monopoly capitalist state unfolds. That sounds like a formula for intensifying crisis and instability, and it is.

CHAPTER 7

Instability and Crisis

THE LIBERAL ECONOMIST'S term for instability and crisis is "business fluctuations," a term that reveals the narrowness of the liberal perspective. It is true that business does go up and down, though it was not until the so-called Keynesian revolution that most conventional economists were willing to consider even *that* an interesting question. It is also true that there are causal links within the economy, that is, among strictly economic variables, whose study can provide some understanding. But for the liberal economist that is the end of the story; such an economist is uninterested in serious study of those aspects of instability that involve political and social links to the economy, or of those aspects of instability that are a product of longer-run political-economic forces. But, of course, that is where the radical's interest begins. The radical economist thinks of "business fluctuations" as something imbedded in a particular social system. One can throw a couple of bottles into the surf and then study the forces that nudge them this way and that, determining their positions relative to one another; a conventional model of fluctuating economic variables might do something such as that. What the radical economist tries to take account of is those towering waves that are tossing *both* bottles about so wildly.[1]

Tendencies Promoting Instability: A Short List

1. A classic example of such a process, operating over the long run, is urbanization and the development of the factory system. The result of this process has been an increasing concentration of workers in urban centers, which in turn makes political organization easier. It also gives the worker some direct insight into the fact that his personal plight is shared by many others, and so into the general nature

of the capitalist system. This process has played a continuing role in the economic history of monopoly capitalism, the post-World War II migration of blacks from country to town being one of its most recent manifestations in the United States. The result, of course, is an intensification of class struggle.

2. The revolutionary technical developments in transport and the media of communication have had a somewhat similar consequence. These are usually discussed in terms of the added power they give to the state to control dissidence. However, the physical mobility and the ability to develop and maintain lines of communication, even among hunted revolutionaries, has been dramatically demonstrated in recent years. The oppressive nature of the world system of monopoly capitalism is now a commonplace, with incident after incident exposed, often naively, in the conventional media and interpreted in the dissident press. As a result, defenders of monopoly capitalism have been forced increasingly into a defensive posture: The tinder has become much dryer.

3. There is a strong secular tendency toward increasing volatility of demand. Investment is, of course, notorious for the volatility of its demand, a consequence of its postponability and the uncertainties of its profitability. But the various categories of waste and surplus in both production and consumption are also volatile, and these are also categories where fashion and caution can produce sharp variations in demand over short periods of time.

4. The internationalization of the capitalist system, combined with media and transport revolutions, has not had the expected effect of averaging out the consequences of local catastrophes on supplies of goods, and especially of raw materials. Instead, the interdependence, combined with very weak mechanisms of coordination, has universalized supply uncertainty. Among the most volatile indexes to be found in the economist's collection is that for commodity prices. Furthermore, the synchronization of crises internationally has led to the recent generation of multiplier effects, so that in general internationalization has tended to intensify instability.

5. As capitalism has developed so have capital markets. Financial markets are larger, more complex and more volatile now than ever before. There is literally no place an investor can go to secure his capital these days, which is a striking contrast with the past.

6. Governments have now become by far the largest organizations in the capitalist world. Consequently, they require expert management to be made to do their repressive jobs effectively. But as was noted in the last chapter, they are so structured that by their very nature they cannot do this job well. It is true that the rather steady increase in government spending has had some stabilizing influence on capitalist markets. But studies have suggested that, for example, in Britain the effect of government policies in the fifties and sixties was generally destabilizing for the domestic economy.[2] And, much more important, government inefficiencies raise serious doubts as to its ability to respond effectively in crises. The mismanagement of the

economy under the Nixon administration, a time of unusually powerful centralization of government authority, is a case in point.

7. The tendency toward stagflation seems to be increasing strongly. The continuing intensification of class struggle has made it increasingly difficult for the usual methods of inflation and recession controls—make the workers and the poor pay—to be implemented.[3] And every proposed policy has differential effects on the various business sectors, so that the conflict among competing capitals further reduces the flexibility of government response. The problem is further complicated by strange alliances, such as those between big business and big labor to protect both labor and capital in a particular industry at the expense of the rest of the economy.

Underlying Forces

The above list is by no means complete, but it is long enough to make the point that there is a considerable variety of factors at work in the capitalist system that push in the direction of increasing instability. There have probably been two main forces pushing in the opposite direction, namely, the rise of government expenditures as a share of total economic activity and the development of monetary and fiscal policies that are designed to counter economic recession. Before appraising the currect balance of forces promoting and inhibiting crisis, however, we need to take a look at the underlying causes of instability, at the basic features of the capitalist system which are fundamentally divisive. Traditionally, radicals have emphasized three crisis-promoting aspects of capitalist relations of production.

Class structure is clearly of central importance. A society that is split into two mutually hostile groups is a good candidate for instability. Each group will try to shift the costs of a deteriorating economic situation onto the other. The more powerful will probably win, and this will deepen the hostility of the less powerful group. And since capitalism has been an unstable but growing society, the losing group is growing in numbers, and also in class consciousness. It is rather natural to expect this situation to produce a tendency toward increasing instability.

One specific way in which class conflict has manifested itself as instability is in generating strong pressures toward an economic downturn. A persistent feature of the later stages of an economic upturn is a more rapid rise of wages than of profits. Labor tends to be at its most militant during such times, because the unemployment rate has been reduced by the earlier part of the upturn, and so the relative demand for labor has become greater. Business fears that continuation of the boom can lead to an actual downturn in profits, and so it cuts back on capital-spending plans. There is also awareness that a mild recession, which brings labor costs back into line, can be the prelude to another high surge in profits. This in turn occurs because of cost-cutting during the earlier phase of the downturn, and because labor can be rehired in such a way as to make maximally efficient use of existing plants. Thus when viewed from the perspective of two or three years, a recession may well look like the best feasible course for business. Not so for labor, of course, and history shows whose interests have tended to dominate.[4]

A traditional feature of economic activity under capitalism has thus been the ability of the ruling class to shift most of the burdens of recession onto the working class. However, this ability has tended to be eroded with the increasing political power of organized labor. The consequence has been not an end to recessions but a cost to both business and labor. Both sides become locked in a struggle to preserve their slice of the pie. The result is more inflation, an erosion of the purchasing power of both profits and wages, and a deeper recession. This process of class struggle has been one of the major causes of the recent stagflation and recession cum inflation.

Another place where class struggle has played a central role in generating instability has, of course, been in international economic activity. The class difference is most striking in this sphere in relations between the Third World and the developed capitalist countries. And the class struggle is fought mainly in the money and commodity markets. The tremendous debt overhang that requires many Third World countries to spend a fifth or more of their export earnings just to pay off the interest obligation gives much political and economic leverage to big capitalists and their governments. But it also raises the threat of bankruptcy, of nonpayment, and these factors in combination enhance the already great instability of capital markets. The intense bargaining over the terms of commodity trade needs no comment. The fact that OPEC could raise the price of crude oil over four-fold in a year is a measure of the extent of the ripoff that this trade has typically involved for developing countries. And the prospective volatility of both price and supply that is suggested by the OPEC experience is an indicator that these always fluctuating markets are about to become much more so.[5]

A second crisis-promoting aspect of capitalist relations of production consists of the effects of the struggle among competing capitals. In times of survival crisis, the capitalists can frequently manage to present a common front against the working class. But at other times their competition among themselves for a bigger share of the surplus pie can be very intense, which is to say destabilizing. Class conflicts in the international sphere are exacerbated by this factor, for all too often capitals there are also capitols, that is, whole national committees of the bourgeoisie become lined up against one another. The pre-World War I rivalries are the classic case in point, but the story today is of continued conflict, though not yet to the point of military confrontation among the major capitalist nations. Instead the battle is fought in international money markets, as nations seek advantageous situations for their exports and protection from threatening imports, and in trade competition, most notably these days in the infamous, government-supported and highly profitable international arms trade.[6]

Competing capitals also have a major role to play in preventing timely and effective domestic intervention by government to reduce the volatility of capitalism. The problem is that some capitalists profit from high interest rates, some from lower ones, some profit from government stimulus to the economy far more than others, some are helped by a general export drive, others are hurt. Compounding this are the many opportunities individual businesses have to profit from particular expansions of government activity—and on occasion contractions. The burden of all these little conflicts is substantially to tie government's hands in many directions that are crucial for economic control. And the general tendency seems to be for more government, even when there is widespread recognition that such policies are likely to be destabilizing.

Finally, there is the so-called anarchy of the market. Marketplace behavior in nineteenth-century capitalism was not only uncontrolled, there was no self-regulating mechanism. Adam Smith's invisible hand referred to what is known as equilibrium, and meant, very roughly, that once all deals that people are willing to make are made, no more deals can be made that make people better off. But neither Smith nor later writers explained how a system of markets could move dynamically, in response to continuing deal making and the changing environment, to produce this state. And there is good reason to believe that the system will not generally tend to settle down in an equilibrium. For one thing, the participants have very little information about what is going on in the economy outside their own petty spheres, and this is clearly not enough to provide an effective guiding hand. Furthermore, expectations of participants are formed on the basis of this limited information, which is obtained from markets that are both out of equilibrium and changing rapidly. That this will produce anarchic results—for example, far too much investment in one industry and far too little in another—is to be expected. And though price changes may eventually tell everyone a mistake was made, it is too late then, and besides, the new prices may very well exaggerate the situation. Anarchy is a very plausible name for the private market system, indicating just how deeply volatility is embedded in it.[7]

For many years now, liberal economics textbooks, almost the only kind available in the United States, have been claiming that the business cycle has been tamed by Keynesian economic principles. That argument has lost some of its cogency with students who have been themselves experiencing the worst recession since Keynes's major work was published. And with good reason. The failure of Keynesian policies has many dimensions, but the most fundamental stems precisely from the anarchy that continues to pervade the capitalist marketplace. Big business does not alter the fact of anarchy, for each of the hundreds of large corporations is itself autonomous and in intense and direct competition with a number of other businesses. Decisions that will affect potential profit levels for years to come have to be made in a highly uncertain environment. Furthermore, the uncertainty is increasing with all the volatilities we have been discussing. And fashion often sweeps up whole rows and banks of executives into a flurry of investment that turns out to be excessive, or of caution that serves only to justify the fears in a new downturn. Even if the government could behave as an effective economic manager, there is little reason to suppose that it could control this anarchy with the limited tools of the Keynesian policy package.[8]

The Structure Crisis

At times the capitalist system has presented the world with a relatively mild and stable surface. From the post-World War II recovery and the Korean war to well into the Vietnam war, developed capitalism seemed to be moving on a fairly smooth track. The great conflict between liberal capitalism and fascism had apparently been resolved in favor of the former. The great imperial rivalries gave way before the hegemony of the United States. And economic growth was moving along at a record pace that was marred by only relatively modest stops and downturns. Government leaders everywhere began talking like Keynesians, and economists began to speak of the possibility of ending recessions forever.

That is just about what one would expect from those bottle theorists. The radical story, of course, was very different. An appraisal of the forces promoting instability shows that an unusual combination of circumstances had produced this apparent harmony; it also shows that the interlude could not last very long, and that it may well have had some of the properties of an Indian summer.[9] Britain, the first great industrial nation, offered an example of the kinds of difficulties associated with inefficient government and class struggle combining to generate intermittent crisis. Conflicts between hard- and soft-currency nations, with West Germany increasingly leading the former group as the United States began to flag, revealed the fragility of the "international monetary order" constructed so painfully in the early postwar years. By the mid-sixties the countries who had carried Keynesian-type planning farthest, such as the Netherlands and Sweden, and other planners such as France, found themselves increasingly unable to effectively control inflation. Third World countries, whose resistance to exploitation was somewhat muted as many of them strove to gain domestic economic control in the aftermath of political decolonialization, increasingly found their international voices. They also found themselves heavily in debt and bound by the monopolistic controls of international big business over their main markets. Expanding socialism was at times a stabilizing threat, serving to unite international capitalists against the greater menace of communism, but as the Soviet Union achieved nuclear parity and then the ability to offer substantial support to national liberation struggles far from its borders, as in Cuba and Vietnam and Angola, a new type of crisis of expectations had made its appearance.

The picture that emerges is one of a crisis that has many dimensions, only some of which are purely economic. The system of monopoly capitalism has shown by its history that it can surmount crises that consist of a combination of one or two of the above factors. But the clouds seem to be darkening and the prospects increasing that a crisis soon will come that combines many, possibly even all of the instability-producing factors at a single point in time. Such a crisis would be unprecedented, and there is no theory to predict for us its consequences or to offer capitalists guidance through it. But it takes no fancy theory to show that most of these crisis-inducing factors are specific to the monopoly capitalist system, and therein lies both an additional threat of crisis and grounds for hope.

CHAPTER 8

Development and Imperialism

THERE ARE fifty or sixty nations around the world, members of the Third World, that, though they are not monopoly capitalist in their political economic structure, nevertheless have basically capitalist institutions. Some of these countries, such as India and Bangla Desh and Chad and Honduras, are desperately poor. Others, such as Taiwan and Greece and Chile and Morocco, are somewhat better off. But all use markets and private ownership of the means of production as basic resource-allocation instruments, and in all of them the state tends to show special favoritism toward the wealthier citizenry. These countries are the subject of this chapter.

One historical point should be made at the outset. In the half century before World War I, every decade or so a new country entered the ranks of the world's great industrial nations. The last country to do this was Japan, and she did it early in the twentieth century. Since World War I not a single country has entered this charmed circle of advanced, industrial, capitalist nations.[1] One of the things that needs to be explained is why there was this dramatic change in the relative progress of capitalist countries. Given the timing, the reader will have surmised that it has something to do with the rise of monopoly capitalism.

Creating Underdevelopment

Developing countries, like their more developed counterparts, have an internal dynamic of their own. In modern times much of this internal dynamic stems from the creation of national markets as a result of transport development and the creation of a central administration through the penetration of the state into the most important areas of economic life. Behind these phenomena are the familiar surplus-extraction processes, adapted to the lower levels of the forces of production in the developing capitalist countries. For example, much of the surplus-extraction operation continues to be by the traditional debt-peonage structure of the countryside; and a good deal more comes from the extractive operations of mines and plantations. Nevertheless, developing countries have their modern enclave as well, and it is here that more modern processes of surplus extraction occur.

Such a description suggests that developing countries are simply relatively poorer versions of developed capitalist lands. But that would be quite misleading. To explain why, we must first move back in history a bit and observe the process of primitive capitalist accumulation at work in these countries. There are famous stories describing the more spectacular instances of this process. The looting of India by the British in the eighteenth century is one of these, resulting in the sending back to England of sufficient wealth to support a fairly substantial part of the early investment in the industrial revolution. Another well-known case is the Opium War and its aftermath in the 1840s, when the British defeated the Chinese in order to prevent the latter from eliminating the flow of British opium into their own country. Once again the profits accruing, after the troops had created the appropriate environment, were very great. Similar if at times less spectacular developments were occurring all over the world after the voyages of discovery.[2]

However, our concern is not with the profits made by the conqueror but with the nature of the losses sustained by these Third World countries. The loss in bullion and luxury craft products was serious but perhaps not fundamental. More important was the effect on both the economy and the social structure of these countries. For example, the continued foreign demands on the Chinese government forced a reorganization of the taxation system in order to meet "indemnity" payments and sharply restricted the government's ability to engage in any productive activities at home. In India preferential rules destroyed the Indian cotton textile industry, substituting factory-made English goods for the products of Indian crafts. Important sectors of the dependent economies tended to be destroyed by various aspects of the country's "contact" with the West.[3]

But the depredations did not stop with economic stagnation and decline. There were also fundamental social effects. The national elite tended to lose its legitimacy, either directly because of the institution of colonial rule or indirectly because of its de facto powerlessness. There was no one able to think through and execute any programs of social or economic development. The internal dynamic of the nation was broken by the conquerors. Nations that had already taken a number of the basic steps that would ultimately lead to modernization were stopped in their tracks by this phenomenon. In this sense underdevelopment is not the result of *being* backward, but the result of a process *imposed* on the Third World countries by the imperialist powers.

How does this history affect the current situation in developing capitalist countries? Of course, technological backwardness and the relatively low level of labor productivity is one consequence. But the current social structure is also deeply affected. The debt-peonage land system is an efficient surplus extractor but is usually a highly inefficient system in terms of the modernity of the production processes it can utilize. This means that there is a substantial social obstacle to bringing about major improvements in this crucial area.

But the problems go still deeper. The colonial or quasi-colonial regime did provide a place for relatively affluent national landowners and merchants. These groups maintained their niches by not rocking the boat, being content with the pickings that were thrown their way. Consequently, even after so stirring an event as decolonialization they continue to represent a powerful braking force on economic development, preferring the certain income from current conditions to the uncertainty that rapid change would be bound to bring. These and other traditionalist groups in a new nation tend to promote a society fractured by differences over the proper policies to adopt.

Such countries will also have development-oriented groups too, people who perceive economic opportunities in the expansion of production facilities. However, their ability to effect change is restricted by the above factors. For example, modernizing agriculture requires a simultaneous development of industry to absorb surplus hands from the countryside; but integrated programs of development are the very things that cannot be achieved in such a brittle political structure. And much of the potential increase in labor productivity that might be achieved is siphoned off by another major interest group thrown up by the distorted social structures of these countries: the military.

Despite all these handicaps, some growth has in fact occurred in a number of Third World capitalist countries. But this growth is strongly conditioned both by the social structure we have just described and by another factor of supreme importance, namely, the role of foreign capital and imperialist policies in the domestic development of these countries. To appraise this phenomenon we must first take a look at contemporary imperialism itself.

Continuing Underdevelopment

The twin destruction of both wealth and social structure is a partial explanation of the failure of any new industrial capitalist nations to emerge in the last sixty years. Handicapped by their past, today's developing capitalist countries start from farther back and have greater difficulty establishing the needed modern institutions. If this were all there were to the story, *contemporary* capitalism would not be responsible for these countries' plights and there would be little that could be done to help them. However, clearly there is more to the contemporary story than this. For there is not only the problem of making up for the past, but also the problem of being pushed still further back by the present.

Another fact that requires explanation is the peculiar impact of technology on developing capitalist countries. Modern technology should bring the cornucopia of higher labor productivity in its train, but instead it seems more often to bring disaster. There is the all too obvious connection between modern health technology and those hundreds of millions of malnourished human beings. Another example is the Green Revolution. This new agricultural technology, combining special seed with irrigation and massive use of fertilizer and pesticide, can bring a doubling and more of output of rice and wheat. However, the poorer farmers either haven't enough land or can't afford the required inputs to take advantage of the new approach; so when increased output—by the richer farmers who *can* afford it—lowers the price of grain, their situation worsens. And, of course, the Green Revolution substantially increased the dependence of agriculture on world market prices. The dramatic oil price increase, acting through the cost of the fertilizer that petroleum is used to produce, priced the Green Revolution out of most markets, leaving the farmers to foot the bill for their investments with no return to show.[4]

Another example comes from the transport and communications revolution, which has been at least partially introduced in developing countries. By increasing mobility and, more important, by introducing the agricultural population to modern ideas, products, and behavior through radio, television, and the like, this revolution has done a good deal to create the culture of poverty, the peculiarly fragmented forms of life and attitudes that are endemic to much of the citizenry of these countries. By destroying the old culture and by emphasizing the relative goods deprivation of the masses, tawdry materialistic values are created that seem to exacerbate the commodity fetishism of traditional capitalism.[5]

Why are these distortions created? Much of the answer lies in the role played by imperialist nations in the social development processes of Third World countries. Over the last quarter of a century imperialism essentially means the United States, though Britain and France and others, including even Portugal, itself an underdeveloped country, have also been in the game. Using the modern instruments of imperialism, developing countries continue to be looted and to have their social structures gutted by outside intervention.[6]

The major instruments of imperialism are foreign trade, foreign investment, foreign aid, political pressure, subversion, military intervention and war. A century ago a favorite practice was conquest and the establishment of a colonial administration. But just as slavery became modernized in ancient times by developing an effective market in slaves, thereby increasing the extraction of surplus, so have the instruments of imperialism become more subtle and effective. Client political regimes, nominally independent, have become the successors of many former colonial administrations, with consequences that have generally been quite satisfactory to monopoly capitalists. Most large multinational corporations seem to generate far higher profit rates abroad than at home. Let us run briefly through the operations of each of these instruments to see how they affect the dependence of developing capitalist countries.

Foreign trade works in a very similar way to debt peonage as a surplus extractor. Developing countries mostly export commodities, minerals, and agricultural products. These are exported in strong competition with other developing-country suppliers, and for the most part are purchased by a highly organized cartellike group of importers, who then further process and market the goods. Such trade tends to produce rather wide fluctuations in incomes to the developing countries; being poor, they must finance their own import program by borrowing when export earnings are down. Thus through both trade and debt a pattern of economic dependence and a basically exploitative relationship is created. It is a sort of international version of the debt-peonage system.

In principle, the foreign trade instrument can be effectively applied without any imperialist presence in the developing countries. Foreign investment, however, is a very different matter. In this case a foreign company, most of the time an American multinational, establishes a subsidiary abroad. This involves importing machinery, employing local workers to construct the factory and to work as its operatives, and brings in further foreign exchange via the foreign managers and technicians who come to live near the factory. This, it is often claimed, stimulates domestic demand and so provides fundamental benefits favoring the creation of a modern industrial base in the developing country.

The true impact of foreign investment is somewhat different. In the first place, one must not forget that foreign investors expect a profit from their operations. This means that some portion of the economic benefits flowing into the country flow back out again later in the form of repatriated profits. It has been estimated, for example, that in Latin America over much of the postwar period there has been net decapitalization; that is, in most years the sum of profits repatriated exceeds the total value of new foreign investment.[7]

Developing countries try to control these losses with the limited instruments available to them. One of these is currency control, used in support of policies aimed at compelling foreign capitalists to reinvest locally some portion of their profits, and to prevent too much of the country's limited exchange earnings to be wasted on luxury-goods imports. The net result of this, given the power and wealth of the opposing interests, has been a massive illegal operation in currency and goods, which in turn means that the above assertion about decapitalization must substantially understate the true losses incurred from foreign investment.

Perhaps even more important is the impact of that new technology the foreign investment brings in. The problem here stems from the divergence between the economic interests of the importing capitalist and the development needs of the society. Much of the new technology will be oriented toward serving the world market, and so the level of operation of the factory will vary with world market profit opportunities rather than domestic needs. The new factory designed to capture the growing domestic demand for a good will probably be given a tariff wall to protect it against competition and so to ensure a monopoly price for the product. And, most important of all, this new technology will have a very high output per worker as compared to earlier times, so that labor absorption will be relatively low. These factors in combination provide a major reason why the real incomes of the poorer half of the population in many fairly rapidly growing economies are actually falling.

Foreign aid, one might think, is unequivocal in its economic impact; surely it is positive. No doubt a $100 million extra in foreign exchange is, other things equal, better than its absence. But that is about as far as one can carry the positive story. On the other side of the ledger is, first of all, the modest and recently declining magnitude of aid. A relatively minor movement in commodity prices is sufficient to cost developing countries a greater foreign exchange loss than the total of foreign aid received in most years. Second, there is its distribution. Most of it has been for military rather than economic "development," and most of it has been allocated, not to the countries that can make the best economic use of the funds, but to countries who happen to be located along the fringe of the socialist camp.

And, of course, the big problem with foreign aid lies in its political implications. Not only does it give bargaining leverage to the donor country, it has substantial impacts on the structure of the domestic country's society. Much of the military aid is not aimed at repelling invasion but rather at putting down incipient revolution. This distorts the internal dynamic of the country. The increasing power and influence of the military has the same effect. The military represents a conservative influence domestically, and is often both politically active and oriented toward modernization. In combination, this has gone a long way to explain the rise of "growth fascism," the military regimes of such places as Brazil, Greece, Bolivia, Taiwan, and elsewhere. Consequently, though foreign aid seems to have had rather little direct impact on economic growth, its political and social effects have been very substantial and very harmful.

The primarily political instruments of imperialism, such as political pressure, subversion, intervention, and war, are spectacular enough to require no description. These are also the traditional instruments of even precapitalist imperialism. The sending of marines into Lebanon or the Dominican Republic gives one a nostalgic sense of continuity with the past—unless one is among the many victims of the act. The CIA's Dirty Tricks Department provides stories to match the bedtime reading of the thriller addict—again, provided one has no direct involvement in the human costs of these operations. The massive slaughter in Vietnam can be matched, and not on a smaller scale either, in many countries whose populations are doomed to extremes of wretchedness as a consequence of the political regimes installed by such interventions. Some of the most appalling examples can be found in American imperialism's back yard, Central America, where infant mortality and illiteracy rates are among the highest in the world.

However, there is an important difference between the behavior of modern monopoly capitalist imperialism and its precapitalist forms. Basically, these direct political interventions represent a failure of monopoly capital, whereas in earlier regimes they were simply the device by which control of a dependent country was obtained. The modern practice is labeled neocolonialism. The idea is to utilize a domestic bourgeoisie to administer the state and a domestic military force to prevent revolution. Monopoly capital then works its magic via the three economic instruments we have just discussed. Marcuse might well call this subdued conquest. When overt acts of force by the imperialist nation are required, this is a sign that the basic neocolonial policy is not working. Of course, the aim of the intervention is to restore it to working order.

Once again though, the irrationality of monopoly capitalism forces a qualification of this distinction between precapitalist and monopoly capitalist imperialism. The military-industrial complex back home feeds on expanding military power. Its legitimacy requires at least occasional use of some of the massive force created over the years. Consequently, premature and irrational application of force in neocolonial areas, even from the point of view of the monopolists' own interests, is also a fact of contemporary imperialist life. Opinions may differ as to whether it was a mistake, from the point of view of effective imperial control of the Dominican Republic, to stage that intervention. To put it more precisely, the Dominican intervention was clearly a plus from the point of view of the segment of capitalists who draw their primary profits directly from the military; what is not clear is whether it was also a plus for the segment of capitalists who expected to profit from their economic operations involving the Dominican Republic. A "mistake" of that kind can be expected from time to time under neocolonial imperialism.

Uneven Development

What are the prospects for the developing capitalist countries that find themselves in this situation? To appraise this question, it will be useful to run briefly through the domestic situation of a typical developing country to get some idea of its potential as well as of the obstacles to realization.

The first thing to notice is that the relative surplus even in the poorer countries is quite large, probably somewhere between a fifth and a third of current output. This is one of the most surprising results of the study of economic development. It was put forward by a Marxist, Paul Baran, twenty years ago, and has since been vindicated, most especially in the Chinese experience. After coming to power in an area containing one of the world's most impoverished populations, the Chinese communists have not only succeeded in delivering basic goods to the whole population, but also in diverting a fifth or more of their output to the expansion of productive capacity. It is now widely accepted that all but a few developing countries possess a relatively large surplus over the basic needs of their populations.[8]

The misery in which much of the population of Third World countries is sunk has already been mentioned several times. But a few remarks about that misery are still in order. First, there is the connection between food and productivity. Many of these people are physically incapable of supporting themselves because of a debilitating diet. More tragically, some portion of the 200 or 300 million children caught up in this wretchedness are being permanently damaged by their current levels of malnutrition. And even those who are capable of work are likely to be substantially underemployed, either on tiny farms or in shantytowns where there is virtually no work to be had beyond the basic struggle for survival in an alien and unproductive environment. Many of these countries not only suffer from all the above problems but also face rapidly growing populations, which year after year are compounding the misery.[9]

All of this can serve as a measure of the irrationality of developing capitalist societies. But that irrationality cannot be fully appraised until the situation of most of the population is placed alongside the enclave development of the so-called modernized sector of the economy. Here in central cities there will be high-rise apartments, fancy cars, stores full of Western goods, and factories turning out the products of contemporary capitalist society. Workers will receive very low wages by developed-country standards but will be substantially better off than the citizenry described above. There will be a fairly substantial set of fashionable neighborhoods that will contain the villas of the national capitalists, the upper-middle class, the professionals, politicians, and military officers, and, of course, the foreigners who occupy the key places in much of the economic activity. In a well-set up country it will be possible for the wives and children to go downtown to shop and to head to the various fashionable recreation spots without ever having to come close to the misery in which they are embedded.

This fundamental social contradiction is not unknown to past history except for one fundamental difference: Over the last ten or fifteen years the bottom half of the population will likely have suffered a fall in real income, while the real incomes of those in the enclave have risen dramatically. This phenomenon is even observable in Brazil, which possesses one of the world's fastest growing enclaves, and in India, which sometimes claims to be socialist, and where the fact that the poorest part of the population is worse off than it was twenty years ago seems inconceivable.

Clearly, countries in this state are ripe for dramatic structural change. One almost wonders how things could have persisted like this for so long. The explanation of this persistence comes in two parts. The first is that extraordinary efforts have been made to preserve the enclaves intact because of their benefits to monopoly capitalism and because the alternative is feared and expected to be communism. It is one of the facts of revolution that the most wretched are not likely to play a leading role simply because their health and energy levels do not permit strenuous action even in their own interests. This tends all too clearly to blunt the enthusiasm of monopoly capitalists to work seriously to eliminate the misery.

Second, the enclave is allowed to reproduce many of the social relations of domestic monopoly capitalism; this means a substantial income differential between those caught up in the enclave and the rest of the population. This aristocracy of dependent capitalism is, of course, the group that possesses the domestically held knowledge and skill needed to run the modern sector of the society. It is also likely to be a group in which the level of alienation is quite high, since many of them have only recently cast off the traditional culture in which they were inculcated in their youth. This group forms a central element of the domestic political support for the existing regime.

In the countryside there is a smaller group whose allegiance also lies in this direction. These are the moneylenders and traders who prey on the peasantry. Their success requires the threat of force to function effectively and so, though they are perhaps typically not really a part of the modern enclave, they tend to support it politically. Then too there are the actual wielders of force in the society, the military and police. These will be trained in modern methods and probably will be fairly well-equipped to carry out their jobs. Once again, the job is protection of the enclave and its supporters from domestic threats.

Finally, lined up firmly with these groups and often orchestrating their joint efforts are the foreign capitalists and their political representatives. They bring with them the experience gained in preserving the system through the more than fifty wars that have marked the Pax Americana of the last quarter of a century. They also can offer outside support at levels essentially restricted only by the level of need.

This panoply of subversion and coercion represents the forces of law and order in developing capitalist society. It cannot succeed in preserving tranquility, given the nature of these societies, but it has so far proved able to suppress revolution with that combination of co-optation and surplus extraction that constitutes its basic modus operandi. Can it be defeated? Before answering that question we must turn to twentieth-century socialism for a look at its achievements, the nature of the alternative it offers developing countries, and the processes by which they might adopt it.

CHAPTER 9

The Rise of Socialism: The Soviet Union

THERE HAVE BEEN three great, innovative, socialist revolutions in the twentieth century, the Russian, the Yugoslav, and the Chinese. Each of these revolutions was great in the sense that it entailed a massive and sustained struggle before the leadership could be brought to effective political power. Each was socialist in the sense that the leadership was committed to the abolition of capitalist relations of production, possessed the will to carry out this massive structural change, and took the workers and poorer peasants as its primary power base. And each was innovative in that in each country a very different set of relations of production has developed out of the demise of capitalism.

There is a very important similarity among these three revolutions. All three were cases of revolution coming to an economically backward country at a time of national disintegration. As a result of this the revolutionary party was able to seize power over a portion of the country, and was then forced to spend years in armed struggle in order to obtain general control of the country. This appears to have been a very vital period, a time of molding of leadership and cadres, of the development of an experienced and disciplined body of men and women who can form the vanguard for the difficult social transitions that must follow the political success of the revolution. A necessity for the survival of the revolution, this period and process has also had a profound effect on the revolutionary process in each of the three countries.

The Russian Revolution [1]

Revolution is a very complex process, a point that is easily missed. After all, there is an overriding purpose, which is simply the seizure of state power, and this requires crude and brutal actions rather than sensitive and complex ones. And even after political success there is still a need to use the cruder forms of coercion against the substantial number of defeated but not yet quiescent enemies of the revolution. Even the economic problems are likely to loom as stark and simple issues: how to feed the population, how to reconstruct a shattered economy. The cruder economic devices of rationing and direct control of key commodities are likely to be at the center of attention.

And yet during these same years in which attention and effort is so sharply focused, a variety of things are happening that will profoundly influence the historical course of the revolution. A brief sketch of this process in the Soviet Union will serve to illustrate the key issues that arise in the creation of a new society.

The first thing to note is that the Russian revolution was history's first successful socialist revolution. There was no relevant experience to draw on. The leadership was without experience in managing either a state or an economy, and what was known of the few weeks of the Paris Commune was of little help. Furthermore, much of even the top leadership of the Bolsheviks had not worked together for any extended period of time, so that even the working understandings had to be acquired very hastily.

Beyond the uniqueness of the experience was the very limited control over events that circumstances permitted. Essentially, the revolution in the countryside ran its course with little more than verbal encouragement from the Bolsheviks. In the major cities things were at least somewhat better; but many factories were in the hands of factory committees who were not all firmly committed to the Bolsheviks. The czarist army had disintegrated, which was a great advantage, but which also provided no organized base from which a new army could be built. And within a few months of October a civil war with the reactionary elements was in full swing, posing a very serious survival threat to the Bolshevik government and demanding all its energies just to keep the troops and supplies flowing to the several fronts.

In February 1917, there were probably no more than a few thousand Bolsheviks. Their numbers, of course, expanded very rapidly after that, and experience in the revolutionary army was a very effective way to build dedicated cadres. But the Russian civil war and accompanying economic chaos were extremely destructive. A large fraction of these cadres had fallen on battlefields and from sickness during the three years of civil war. The survivors were far too few to handle the massive administrative tasks of operating the Russian state and economy during the twenties; furthermore, their war experience had not provided them with any special understanding of how to organize a socialism that would satisfy the peacetime needs of worker and poorer peasant.

Unfortunately, the leadership had learned perhaps more about how not to operate a peacetime socialist economy than about how it ought to be done. The high degree of paper centralization of the civil war economy, the so-called war communism, was in fact little more than an ad hoc requisitioning of the most desperately needed goods and their intermittent delivery to the most desperately needy potential consumers. Clearly, it was no way to operate a peacetime economy, and it had already aroused considerable suspicion of the Bolsheviks in the countryside. The retreat to decentralized market forms of operation of much of the economy was clearly a response to these factors.

Nevertheless, the state still had immense requirements for cadres, and these had to be filled from the ranks of careerists, opportunists, and Old Bolsheviks, the latter in a rather small minority. A periodic weeding out of undesirables was no doubt helpful, but the Communist party could not be a terribly effective instrument of policy, especially in areas where strong commitment to the ideals of socialism was needed. And it had no deep roots in the countryside, where most Soviet citizens lived, and where the revolution had, without Lenin's ability to affect the outcome, thrown up a system of peasant freehold agriculture whose relations of production were basically incompatible with socialism.

A final factor needs to be mentioned. When the revolution did not materialize in the rest of Europe, the Soviets were forced to accept the environment of "socialism in one country." This was a hostile international environment, one in which continuing attempts to frustrate and bring down the Soviet government were made. It was an environment in which trade was extremely limited and loans nonexistent. The Soviet efforts at reconstruction and socialist construction had to bear the direct costs of this isolation, which were substantial. They also had to bear the indirect costs, in the form of relative isolation from the worlds of science, technology, and art, as well as from social and political contacts as they were developing in the rest of the world; for many leading segments of Soviet society, this may have imposed an even greater cost.

Achievements of Soviet Socialism [2]

These days it is difficult to appreciate the spirit of creativity that permeated Soviet social life during the twenties. Just a few points will illustrate how far this spirit went and how close was the association between thought and action. In the West the women's movement reached one of its peaks around the time of the first world war, receding almost everywhere after women obtained the right to vote—and after the men returned from war. In the Soviet Union most of the social disabilities that continued to balk the liberation of women elsewhere down to our own time, such as ease of obtaining a divorce, equal legal status in handling finances, the right to abortion, and equal pay for equal work, became official policy and had dramatic effects. In education, curricula were reformed to reflect the needs for vocationally and professionally qualified workers rather than supplying essentially no more than the cultural fare of the gentleman. A statistical system was established that may well have been the world's finest of its time, and one strongly oriented toward supplying data of the kind needed by policymakers. A social welfare system was developed that was oriented toward providing a more realistic system of protection of the urban worker's economic rights than was available in far more affluent capitalist societies. And the world's first full-scale reorganization of the penal system to emphasize rehabilitation rather than punishment was instituted.

Of central importance for the survival of socialism, the problem of management of the economy was slowly being mastered. This was a very creative period in Soviet economics; indeed, in the twenties Moscow and Leningrad were probably the world's most exciting places for an economist to be; and because of the close connection between words and deeds in Soviet intellectual life, one was expected to assume responsibility for the practical consequences of one's academic work. In this environment the tools were being fashioned that would later be adopted around the world wherever at least the semblance of economic planning was practiced. Through the so-called control figures the basic ideas of managment of a relatively decentralized socialist economy were developed in practice. Through the emerging planning bodies the role of the medium-term plan in guiding the economic trajectory of a socialist economy was being developed and, in the late twenties, applied. And throughout this period cadres were acquiring the experience that would be passed on, not so much in manuals as by word of mouth and demonstration, to future generations of planners.[3]

The systematic application of science, technology, and economics in the service of a deliberate process of economic modernization is an invention of Soviet socialism. Both the economic and the political tools for this effort were developed in the Soviet Union. Economists had heard of economic development before, and some governments had engaged in piecemeal efforts to build up the national economy. But never before had an attempt been made to bring the full panoply of human knowledge, skill, and effort to bear on the problem. The result was one of history's great accomplishments: The swift reconstruction of a shattered economy that in 1920 was operating at about a fifth of its level of five years earlier, and the transformation in the following decade of the Soviet Union into the world's second most powerful industrial nation. Economists now realize that this transformation cannot be effectively measured in conventional statistics; essentially, it was a dramatic qualitative change in the nature of the heart of the Soviet economy, and it remains an effort that is probably unmatched in the capitalist world.

The demonstration effect of the existence and behavior of the Soviet Union on monopoly capitalism is very hard to appraise, except for the claim that it was extremely important. It manifested itself first in the wave of revolutionary activity just after World War I, which brought those distant events home in a very striking way. Probably the Soviet Union deserves a great deal of credit for the social welfare legislation that became a ubiquitous part of monopoly capitalism during the interwar period. And the Communist parties that were created around the world exerted a profound effect on domestic developments in tens of countries; their impact on the rise of the independence movements in the Third World is still unappraised but was clearly substantial. Also, the core of militants communism supplied to working-class movements everywhere no doubt had a profound if often indirect influence on the behavior of union and political leaderships, in addition to the impact of their dedication on other workers.

But the Soviet Union had a much more direct impact for several decades as the only major defender of socialist interests in the international arena. That American policy for a few decades was forced into the relatively defensive stance suggested by the word "containment" is attributable almost solely to the existence of the Soviet Union. Socialism in Eastern Europe and Cuba clearly owe their continued existence to the support of the Soviet Union. And adventurism by monopoly capitalism in other areas, from Egypt and the Near East to China, has clearly been inhibited by its presence. Developing the ability to carry out this policy has taken a heavy toll in terms of further development of socialism within the Soviet Union; but it is a central socialist achievement that should not go unnoticed.

One final Soviet achievement should be mentioned, and this may well be the most important of all. It was in the Russian revolution that the great political and social instrument of socialist revolution was forged, namely, the Communist party. This great vehicle for mobilizing whole nations for the task of social transformation is a distinctive feature of all successful socialist countries; and where socialism's success has been dubious, a weakness in the party tends to loom large among the causes. It takes strong medicine to overcome the tensions of fragmentation and wretchedness that abound in underdeveloped countries, and the process of change is fraught with conflict and continued attempts at subversion from both domestic and foreign enemies. One of the lessons of the twentieth century is that a Communist party, organized along the lines pioneered in the Soviet Union, can serve successfully as that strong medicine.

Soviet Failures

Collectivization was one of the most serious, as well as one of the most misunderstood, of Soviet failures. That it was a failure can hardly be doubted. In the first place, there was the terrible human cost, which must run to millions of unnecessary deaths, many of them poor peasants, the supposed clients of a workers' and peasants' state. These deaths came primarily because collectivization was imposed on an unwilling freehold peasantry who had not been brought into serious and trusting contact with the communists. Collectivization appeared to them as confiscation. It turned out in fact to be a regime that supported a brutal and exploitative extraction of their surplus, and even more than the surplus. In fact, Soviet collective farms looked like such effective surplus extractors that the invading Nazi armies decided to retain them in the Ukraine.

Political neglect of the countryside was combined with economic neglect, so that Soviet agriculture has remained a relatively ineffectual part of the economy down to the present day. The peasantry is now far better off than during the thirties and has been the beneficiary of substantial provision of communal services and a guaranteed base income to families; but the inefficiency remains as, in all likelihood, does the unpopularity.

These undoubted failures have often been attributed to all forms of collectivization, and therein lies the misunderstanding. The more gradual and humane processes of collectivization adopted in other socialist countries have produced far different results. One of the most interesting tests came in 1968 when Czechoslovak peasants, had they wished, could have decollectivized and divided up the land once again into family farms. Essentially no collective farmer took advantage of this de facto opportunity, which is quite a decisive test of popularity. In other countries the basic efficiency of the form seems established, and in some places, such as China, the results achieved by collectivized agriculture can only be called spectacular. Building on Soviet experience, later socialist collectivizations have managed to avoid the most serious of the Soviet mistakes in this crucial area.

A second basic failure of the Soviet Union is represented by Stalinism. As a consequence of the brutality of Stalin's rule, from collectivization in the late twenties to the "doctors' plot" on the eve of his death, the Soviet regime was marked by an extreme of political centralization and terrorism that is one of history's darkest pages. Part of this result was simply bad luck, namely, Lenin's early incapacitation and death and the consequent failure of the Bolsheviks to establish an effective succession policy. Part was a consequence of their failure to reach the peasants politically and ideologically. This inserted an element of fear into Bolshevik attitudes and of coercion into their policies, so that such a man as Stalin could seem to many effectively to meet the needs of socialist construction. Stalinism was in part a consequence of the extremely threatening international environment in which the Soviet Union was forced to make its way, an environment so fraught with hate and fear that paranoia seemed almost to be a rational reaction. And Stalinism was, of course, partly a product of Stalin's personality, the tendencies of which had not been fully revealed by the time he acquired power.

The environment of the time called for a very strong central government both for domestic transformation and for protection against invasion and subversion. Stalinism was not exactly an ideal answer, but there is little in the later history of socialism to suggest that he is at all a likely phenomenon under socialism. The Soviet Union has been institutionalizing a regular succession routine in which, roughly speaking, the Central Committee of the party serves as the basic constitutional organ of leadership selection. Other socialist countries also generally reveal a more flexible and open attitude toward constructive socialist dissent than would have been feasible under Stalin. This political loosening up has not gone as far as many would like, but it *has* made Stalinism a thing of the past.

A third area of disappointment with the Soviet Union as the world's first socialist state has been the undeniable manifestations of imperialist policies. The suppression of the Hungarian and Czechoslovak revolutions and the continued intervention in the domestic affairs of East European nations are obvious enough. Some of this can be attributed to defense needs of the Soviet Union, but there is obviously more to it than that. And no one would argue that socialism can develop very far while its domestic representatives can only stay in power because of the threat of foreign intervention.

The extent to which these policies are based on genuine fear of the West may be tested if a substantial troop withdrawal from Central Europe is negotiated with the United States. But there are already signs of a lessening in the use of imperialist coercion through the much greater flexibility in domestic policymaking that has been occurring in Eastern Europe. The Hungarians are continuing their ten-year-old experiment with a form of decentralized socialism, and a variety of new policies have been introduced independently in Poland in recent years. In fact, in these countries and in East Germany some of the more interesting experiments in developing socialist experience with alternative economic structures are being tried. Once again the trends suggest that there is a good deal of learning from past mistakes going on.

But the most important point about Soviet imperialism comes from a much different direction. The Soviet Union's "satellites" are put through the same social revolution as the Soviets themselves were. As a result there is an incomparable difference between life for an average citizen in, say, Bulgaria as compared with her opposite number in a "free world" dependent state such as Nicaragua. The former citizen is hard at work transforming her country, assured of a good education for his children, high-quality health care, and an increasing flow of consumer goods for her family. The latter is probably a hungry peasant whose malnourished children are unlikely to see the inside of a schoolhouse for more than a very few years and whose health will not permit much learning. For the ordinary citizen the difference is between hope and hopelessness. Imperialism is perhaps not such a good word to use for *both* these phenomena.[4]

Conclusion

The Soviet Union was at one time the hope of the world for socialists. That is no longer the case. If socialism were to come to the rest of the world tomorrow, the Soviet Union would have a good deal of catching up to do. However, the Soviet contribution to the development of socialism in the twentieth century has been very great indeed and, as we have seen, there are strong signs that some of that catching up is already occurring. We will return to the issue of reformability of the Soviet Union in chapter 10, but enough has already been said to suggest that the anomalies of Soviet behavior are likely to be more easily corrected than those of monopoly capitalism.

CHAPTER 10

The Rise of Socialism: Yugoslavia and China

THE ARGUMENT that there are many paths to socialism is based on the thesis that there are powerful long-run forces at work in a society conditioning any change that takes place. The forces-of-production concept stands for one such set, for those forces related to the productivity of labor, and in the twentieth century we have been learning a good deal about the amount and rate of change in the forces of production that is feasible. The relations-of-production concept stands for another set, those related to the extraction and allocation of surplus, and we have been learning more about changing them too, though it is inevitably a more difficult concept to come to terms with empirically. But it does seem that the unique histories of countries set different combinations of factors to work conditioning changes in both the forces and relations of production. For that reason it is not really sensible to explore the nature of socialism in a very abstract way.

In this chapter we deal with the aftermaths of two successful socialist revolutions. They suggest the great range of socialist ideas as to what a well-functioning society should look like. Neither, of course, is perfect, but both have succeeded in carrying out fundamental changes in both the forces and relations of production. The idea is to compare very briefly the approach in each country to four basic aspects of society: economic development, participation, decentralization, and the transformation of man. The aim is not to choose between paths; the idea is rather to suggest that on present evidence it seems that a well-developed socialism might contain societies practicing either or both of these approaches.[1]

Economic Development

For the first two decades after World War II, the Yugoslavs followed the Soviet strategy for economic development rather closely. Essentially, this involves a very high rate of investment, up to a third of national income being allocated to increasing the stock of means of production. It also means placing central emphasis on industry in general and heavy industry in particular. The aim, obviously, is the swiftest possible development of the forces of production to modern levels of labor productivity. Despite a four-year depression from 1948 to 1952 brought on by the Cominform boycott, the Yugoslavs followed this strategy successfully, achieving one of the world's highest growth rates of output during this period.[2]

Any strategy involving a concentration of efforts inevitably implies that some things will be left behind. That was certainly true of Yugoslavia. For almost a decade after the completion of postwar reconstruction, there was very little rise in levels of consumption. Of course, communal services such as health delivery and education expanded extremely rapidly during this period, but provision of the marketed type of consumer goods did not expand much more rapidly than population, and the service sector of the urban economy was neglected.

However, the main sector that was left behind was agriculture. At first this represented a deliberate but mistaken developmental strategy, based on the feeling that agriculture would quickly reattain its prewar position as a major earner of foreign exchange. When this did not materialize, more resources were devoted to the sector and a substantial expansion of output was achieved. However, the retention of small-scale peasant freehold land tenure, while providing a basic level of security to peasant families, did nothing to solve the social problems of the countryside or to introduce socialism into it.

Yugoslavia faced a difficult dilemma with respect to collectivization. The problem was that the initial drive coincided both with two very bad harvests and with the Cominform blockade. The need for national unity and for development made a stabilization of the rural relations of production essential. But to go forward would then have meant rapid collectivization à la Russe. Unwilling to take that drastic and inhumane step, the other recourse was to fall back, allow peasants to leave already established collectives if they wished (90 percent of them did so), and to provide security of family farm tenure. This they did, leaving a legacy of tiny farms and much technical backwardness in farming methods.

By the time collectivization could again be contemplated, the Yugoslavs had essentially decided on a different strategy for transforming rural society, namely, migration from the overpopulated countryside. This strategy has worked, in the sense that the peasantry has fallen from two-thirds to one-third of the population, mostly as a result of migration to Yugoslav cities, partly to Western European ones. Collectives may yet become the basic agricultural organization in Yugoslavia, but if they do it will probably be by a process of emergence through felt economic need for larger-scale production.

The lesser social upheaval entailed by the Yugoslav rural strategy has its appeal, but for many countries the Yugoslav approach is simply infeasible. Usually there are few opportunities for migration abroad and the domestic cities have no hope of absorbing the surplus rural population. China, still over three-fourths peasant, is one of these countries, and so the Chinese had to devise a different strategy.

The failure of economic development to relieve misery in capitalist countries occurs precisely in those countries in China's, not Yugoslavia's, situation. The Chinese have, in effect, suggested that the fundamental problem of economic development needs to be reappraised in the light of this terrible fact. Instead of being modernization, the fundamental task of economic development should be to steady-state a humane agrarian regime. This regime must be capable of sustaining the peasantry at a decent level of life during the two or three generations in which massive absorption of the peasantry into developing-country cities will not be feasible. Through their commune system the Chinese seem to have done just that.[3]

The commune—and its equally central smaller units, the brigade and team—provides basic social security for income and health to its members. It is also a cultural and administrative center for organization of nonroutine projects, such as water conservancy measures and other forms of local investment. The lands are farmed jointly by teams, and rewards are related to effort and, to a variable extent, to work attitudes. Each family typically has a small private plot and the right to dispose of its output. Small-scale industry is often developed within the commune. A substantial portion of the surplus is siphoned off by the state, partly by compulsory sales at low prices and partly by taxes, but a large—if unknown—portion remains at the commune's disposal. Clearly, this system has provided for the basic needs of the rural population and appears to be very popular. It differs from Eastern European collectives more in the successful integration of members into the system, and of the commune into the larger society, than in its basic organization.

Of course, the Chinese have not neglected modernization. They appear to have made very substantial strides in this direction as far as urban industry is concerned. But their performance in this area does not seem to be as outstanding, relative to other countries, as in their method of dealing with agrarian misery. It is this latter respect, the acceptance of the peasant as a genuinely equal partner to the urban worker in socialist society, that represents the most significant novelty in the Maoist development strategy.

Participation

Participation means playing an active role in making the decisions that affect one's life. The emphasis is on the word active. For example, voting in a capitalist democracy is *not* participation. The individual cannot affect the outcomes by casting his vote one way or another, and besides, the system is rigged in such a way that money counts for more than votes in producing government policies. By and large participation implies activity that has a significant impact both on the formulation of policy alternatives and on the choice among alternative policies. The mechanisms of contemporary monopoly capitalism simply do not fill this bill.

In China participation probably reaches its peak in the countryside. The peasant understands his production environment; he is even able to appraise many aspects of the applicability of new technology at least as well as a scientist. And he is given some control over that environment, partly through his private plot, partly through the relatively small team that is the basic socialist production unit. We cannot obtain a clear picture of how "leading" a role cadres play at this level of decision making in China, but it does appear that there is a substantial role for active participation by ordinary team, brigade, and perhaps even commune members in the decision process at these various levels.[4]

In the urban factory there also appear to be serious attempts to enlist participation by ordinary bench workers in dealing with relevant infra-factory decisions. Once again there is very little evidence based on serious and extended study of behavior in these Chinese environments, but efforts to improve communication all along the line, and particularly between the experts and the people, are certainly substantial.

However, the limits to participation in China must also be mentioned. There are sowing plans laid down by higher authorities, which are a substantial intervention in the daily lives of peasant producers. The commune and lower units have very limited ability to control the destination of the substantial portion of their output that leaves the farm. Urban workers do not decide what or how much to produce in their factory, nor are they involved in any serious participation in determining the terms on which the goods are delivered to society. The Chinese environment has a very strictly controlled information system, one that must substantially restrict worker and peasant understanding of the environment in which their production unit is embedded. State control of communication also serves to protect the entire system from serious study by outsiders, inevitably leaving open the question as to the relative de facto influence of workers and cadres. There is more participation in China than in large production units under monopoly capitalism, but much of what really happens remains very cloudy.

Yugoslavia has a very different approach to participation. Basically, the Yugoslav belief is that participation is a matter of power. Consequently, their hopes for developing a participatory society hinge on giving the workers in socialist production units the power to make the basic decisions of an enterprise. An elected committee, the workers' council, consisting mostly of production workers, not technicians and managers, is empowered to make such decisions as what to produce, how much to produce, where and on what terms to sell the output and buy the inputs, and, aside from taxes, how to distribute enterprise income among the producers and between wages and investment. These bodies have seen their de facto power increase rather steadily since the system was first inaugurated twenty-five years ago, and today they clearly are a major power in economic decision making in Yugoslavia.

There are limits to that power. The manager remains a figure with dual loyalties, both to the workers' council and to the government. Party and trade unions exert some influence within the firm, though far less than in any other socialist society. And the local government is a power to be reckoned with by most enterprises. Finally, the financial environment, including the (worker-managed) socialist banks, are able to influence enterprise policy. In Yugoslavia as elsewhere in the modern world, decision making is complex, but at every major stage direct representatives of the workers will be found actively participating in the decision process.[5]

Yugoslavs have communes too. That is the name they apply to local government, whose average population is perhaps a half dozen times larger than that of the typical Chinese commune. Here Communist party influence is relatively stronger than in the firm, but citizens selected in genuinely competitive elections also play a major role in decision making. And the communes are rather more powerful than their counterparts in either China or the West in that they are the beneficiaries of a revenue-sharing scheme that gives them a genuinely autonomous tax base, the central ingredient in participatory economic decision making within government. Once again participation on issues of central importance to commune members' lives is likely to be substantial.

Decentralization

There are only two basic forms of interproduction unit relations known in industrial societies: markets and hierarchies. It appears that all socialist societies except Hungary and Yugoslavia have chosen the latter form for their modern, larger-scale industry. Hungary and Yugoslavia are the socialist experimenters with market forms.

When Yugoslavia first adopted the market for relations between production units, it was almost universally criticized by socialists. However, since the period of economic reform in Eastern Europe, and with the growing appreciation among socialists of the alternatives, the market has become a bit more acceptable as a socialist resource-allocation device. Within the socialist sector, Yugoslavia has remained one of the world's most egalitarian societies despite use of the market for almost a quarter of a century now. And both the growth experience and the competitive quality of its goods suggest that the market has served reasonably effectively as a socialist resource-allocation device. Hungary has less than a decade of experience with the market, but there too it appears to be working successfully.

However, all this is not to deny the success of those socialist countries that have continued to use hierarchic systems of relationships among enterprises. This is a form that differs more substantially from monopoly capitalist practice than the market and so one might expect that a relatively long time period will be required to develop its full potential. But probably its greatest comparative advantage lies in its use as a mobilizer of resources in the early stages of modernization. Such a period of crash transformation of the economy bears some resemblance to the problems of wartime economy, where hierarchic controls tend to be ubiquitous.

Both systems offer substantial opportunities for preserving relative income equality, as the above facts suggest. However, it cannot be denied that decentralization produces some inequality, since localized units tend to become residual recipients of windfall gains and of natural advantages under decentralization. Yet it appears that somewhat similar processes occur in bureaucracies as well, screened by the relatively closed information processes of this form.

Emphasis is often placed on decentralization in the Chinese experiment. But the size of the country must also be borne in mind. Szechuan, just one province, has a substantially larger population than France or West Germany. A substantial decentralization "to the provincial level" means in China decentralization to the size of a typical large nation. Furthermore, the centralization of most of large-scale modern industry at levels equivalent to that of the state elsewhere is a fact of the Chinese way.

Some decentralization has occurred in China via the emphasis on development of local industry at the county and commune level. This has become important in permitting relative autonomy for these smaller units of government and production, and is also an important part of the Chinese way. And there has been some development of processes of contracting that allow limited leeway to local bodies in negotiation of the terms of exchanges; however, the extent and significance of these latter is unclear at the moment.

A major criticism of the use of markets under socialism is the same as that under capitalism, namely, its anarchic action. The Yugoslav economy has not been a terribly stable one and has had a recurrent problem with inflation and with a stop-go pattern of output aceleration and deceleration associated with balance of payments difficulties. However, a good deal of this is capitalism's instability being transferred via the world market to the Yugoslav economy. Presumably, a socialist market system has two advantages over its capitalist counterpart: (1) there is no capital market, one of the most volatile and central in the capitalist economy; and (2) the absence of the basic class division between workers and capitalists makes effective stabilization policy much easier to institute at the political level. It is not really possible to separate the relative strengths of these factors on present evidence. However, it has also turned out that hierarchic socialist economies are subject to rather greater fluctuations in annual rates of growth than earlier socialists had anticipated.

Other things equal, socialists will certainly prefer decentralized to centralized forms of resource allocation, so as to permit a maximum of participation. However, as the above remarks suggest, other things are not equal and a variety of considerations may decree relatively centralized forms for most socialist economies, at least for a while. We will return to this issue in the next chapter.

Transformation of Man

A central tenet of socialist belief has always been that man's behavior is strongly influenced by her environment. Give him a more humane environment to live in than capitalism, and human nature itself will appear to have been transformed. One of the principal tests socialist visitors apply to reputedly socialist countries is, Do you see evidence that a process of transformation of man is underway?

Socialist visitors to China tend to come away answering this question with an unequivocal yes, as indeed do many nonsocialists. The sense of solidarity, of working for the good of the society rather than for selfish aims, is very strongly conveyed, even though the barriers of language and culture are exceptionally high in the Chinese case. Certainly there is no doubt of a massive effort by cadres and media to promulgate socialist habits of thought and behavior. And major political events, such as the cultural revolution, seem aimed at continuing this process. Efforts to reduce the status distinctions among various types of work, including especially the requirement that officials spend a good deal of time working alongside the people, are also an extraordinary manifestation of this effort.[6]

The socialist visitor to Yugoslavia will most likely answer this key question in the negative. He will find rather limited efforts at best of the kind described in the last paragraph. When asked about this issue, a Yugoslav defender might reply: We too expect a transformation of man, but we expect it to emerge from unalienated relations of production, not from artificial efforts at indoctrination. The Yugoslav way to the transformation of man is to make the Yugoslav worker the master of her fate, insofar as that can be done in the modern world. Working together in a group that has genuine power over its workplace, and living in an environment in which home life, too, is participatory, we will let human nature speak for itself.[7]

The retention of a substantial private peasant sector and of a significant small-scale urban private sector no doubt sets some additional barriers to this emergence in Yugoslav society. But it is also true that much of the solidarity the casual visitor perceives in China is the result of a highly centralized control of public media of communication. Only a much deeper penetration into the real relations of production in daily Chinese life will permit a judgment as to how effective the transformation of the Chinese citizen has become.

Conclusion

Clearly, the Yugoslavs and the Chinese have very different fundamental ideas as to what developed socialism should look like. In neither case has the ideal as yet been closely approached. But in both cases impressive transformations of the relations of production away from capitalism have been achieved. Some of the differences between the two conceptions of socialism grow out of their differing situations; the Chinese did not have the Yugoslav development strategy open to them. Central to the differences are a different conception of the nature of human nature in society. The Yugoslavs are closer to the anarchist tradition, with its emphasis on voluntarism and the autonomy of the individual in an environment of equality as the basis for self-realization. The Chinese are closer to the solidarist notion that one person's nature is joint with another's so that self-realization is primarily through interaction. Perhaps the appropriate way to resolve this difference is to watch and support these two attempts at creating a new society as they unfold.

CHAPTER 11

Transitions

SO FAR in this book there has been a good deal of talk about capitalism, and a bit about socialism, but none about how to get from one to the other. That is the topic of the present chapter, which considers separately the three rather different transitions: from developing capitalism to socialism, from monopoly capitalism to socialism, and from present-day socialism to real socialism.[1]

Peaceful Transition?

Can there be a peaceful transition from developing capitalism to socialism? No, there cannot. One can be quite definite on this point because of a recent and decisive test, namely, the case of Allende's Chile. Here was a country with perhaps the strongest and longest democratic tradition in the developing world. Chile is a relatively rich country, with much copper and good agricultural potential. Only a fifth of its population were engaged in agriculture, so the special problems of the countryside were not the crucial ones. The Allende government stayed within the constitution, and a portion of the armed forces, led by the now-martyred General Prats, was willing to support Allende so long as he remained constitution-respecting in his behavior. And finally, Allende came to only limited power, holding a minority of seats in the parliament, and so could effect only a sharply limited set of reforms; a full referendum would be held at the next election, which could send liberals back into power if a majority preferred.

It would be hard to design an environment more favorable to a peaceful transition to socialism than this. And it failed. The reasons for failure are simple enough. Domestically, the problem was that there was a set of parliamentary elections three years after Allende's accession to the presidency. Contrary to the constant complaints that were being voiced everywhere in the liberal world about Allende's failure, in these elections his supporters substantially increased their share of the vote. They did not achieve a majority but the threat was now clear: Allende's policies of doing something about poverty and unemployment and economic stagnation were popular. The results of this election brought the still essentially liberal or conservative armed forces onto the scene as Chile's rescuers from the will of the people.

But, of course, action to remove Allende began well before this, and not so much in Santiago as in Washington. The sorry list of measures taken by the United States, from the ITT offers to engage in subversion to the Kissinger decision to mobilize international capital in a boycott of Chile, need not be retold. Obviously, the real list is much longer than the one publicly available. But the latter is long enough to demonstrate that substantial and powerful efforts began to be made as soon as the election results put Allende in the presidency. These efforts were directed singlemindedly toward returning Chile to the "free world."

Had the story gone the other way in Chile, the issue of peaceful transition would still be controversial. For nowhere else in the developing capitalist world is there a country that could come close to matching the traits that made peaceful transition seem possible there. The failure of the Chilean transition is decisive.

At the present writing there is a great deal of talk about Eurocommunism, which many think offers a new opportunity for a peaceful transition in France or Italy or perhaps elsewhere in southern Europe. Certainly the possibility exists for a Communist party to obtain enough electoral votes to be legally empowered to form a parliamentary government. The Chilean case finds a full parallel to this point. But why, one might ask, should the rest of the Chilean script be played out once again?

Of course it need not be, at least not fully. Should a genuinely socialist party find itself in the above position, it will soon have to face the choice between becoming either a liberal, social democratic representative of the left in parliament, or of moving on to genuine socialism. The capitalists, and the history of class struggle in the country, will force that choice. We have seen many examples of one side of that choice in the twentieth century. The German social democrats of Weimar Germany are the archetypal example. The social democrats were the great revolutionary party of continental Europe, inheritors of the Marxian tradition, and after coming to power in Germany in the nineteen twenties by a similar electoral decision, they decided on the course of compromise and accommodation. This gave the capitalists time to regroup, to form political alliances, and to preserve and strengthen their control over the armed forces, and in particular the officer corps. Then if electoral politics did not suffice to eliminate genuine socialists from the seats of power, other means would be available. Furthermore, the environment of parliamentary compromise, acted out in the context of an existing regime of private property relations, tended to sap the socialist orientation of many leaders of the social democrats. Whatever they may have been originally, long before losing power social democrats had become liberals. This path is well worn, the forces that push members of the left down it are very strong, and the consequences are clear: a return to liberal parliamentarism under capitalist relations of production.

But there is another path, one that requires the elimination of the power base of the capitalists. That can only be done one way, namely, by nationalizing the means of production that provide that power base, and particularly the commanding heights of large-scale industry. Once that step is taken the door is open to movement toward genuine socialism. But it is hard to imagine a capitalist class sitting quietly while such a vote is taken, in parliament or elsewhere. This key step will be strongly resisted, will promote a great political, social, and economic crisis. The usual term for such a crisis is revolution, and it will not be resolved in favor of socialism without some elements of violence. But it is a path down which Eurocommunism *may* move, if its leaders still have the will to create a socialist society.

Revolution

A considerable number of reactionary governments have been overthrown from the left since World War II. Among them, in addition to Yugoslavia and China, have been Cuba, Vietnam, Bolivia (1952), Peru, Egypt, Algeria, Syria, Iraq, and, most recently, Portuguese Africa. The overthrown governments in almost every case were in a state of great weakness or even disintegration. The exceptions are the colonial governments, where an essential ingredient in revolutionary success has been, not so much the defeat of the colonial government and army, but the raising of the cost of continuing to operate the colonial government to the point at which the metropolitan government finds independence the less unattractive alternative. The frequency of these occurrences suggests some measure of optimism is justified that we have not seen the last of them.

But the political revolution is only the first long step toward establishing socialism. There is still—for those who made the political revolution—the wholly new set of tasks posed by the morrow of the revolution. Effective soldiers cannot be turned overnight into effective economic managers; nor do drill masters understand the way to turn workers and peasants into convinced supporters of a socialist regime. Something else is needed.

The leading proposal in the field to serve as that "something else" is the Leninist party. It has been the major tool for the mobilization of society for the great tasks of the transition in at least thirteen socialist countries. These thirteen countries have a successful record both in economic construction, in eliminating the basic institutions of capitalism, and in successfully resisting challenges from the right to restore those institutions. Of the countries that have had a political revolution but have not had a Leninist party in the central place as a mobilizing agent, perhaps only Cuba has as good a record as these thirteen; and the Cubans within a very few years of the political revolution were in fact making central use of a Leninist party. The association is too strong to be accidental.

The revolutions in China and Yugoslavia stand out because of the cadre-building that went on during the period of revolutionary struggle. On the morrow of those revolutions there was in each country a large body of men and women whose commitment to socialism was buttressed both in battle and in study of the basic ideas of socialist thought. In both cases these cadres played a fundamental role in the swift restructuring of society and have since, even in Yugoslavia, been essential in sustaining a basically socialist structure for society. As noted, in Cuba the melding of the 26th of July people with the regular party people was carried out over the years after the revolution, but no student of recent Cuban history doubts the importance of these cadres in preserving the revolution. Even in Eastern Europe, where the party cadres often appeared to the population as alien agents, they were central to the elimination of the capitalist class from positions of power and to the creation and operation of the basic socialist institutions.

This is in sharp contrast to the situation elsewhere. Not only does mobilization of the population seem weak in the years after the political revolution in countries such as Egypt, Bolivia, Algeria, and Syria, but economic development seems on the whole rather less successful, and the slipping back of institutions into quasi-capitalist forms also seems a serious threat, not to mention the political-military weakness that made a countercoup successful in Bolivia (Egypt too?). In this list one should be careful not to underestimate the importance of economic transformation of the country; economic backwardness is one of the major threats to the survival of socialism in developing countries.

There is another factor that suggests the strong-medicine approach represented by the Leninist party is essential. This is the reaction of monopoly capitalism to the threat of revolution. The inherently unstable situations of the period of elimination of political colonialism are now largely past, and regimes, often strongly authoritarian ones, are in place and functioning as agents of the new, "nationalistic" neocolonialism. These governments are somewhat less frequently ripe for revolution than was the case in the decolonialization period. Furthermore, the metropolis has devoted great efforts to the development of "counterinsurgency" techniques, from helicopters to "scientific terror," including torture. And the imperial metropolis is prepared to spend a great deal of money and effort in training and equipping local counterinsurgency groups. This does not by any means end the possibility of revolution; but it suggests that the struggles are not going to become easier in coming years. What is basically needed are trained and committed cadres who are ready to seize the time when it comes.

Monopoly Capitalism

There are two important differences between developing and monopoly capitalism that argue for a different approach to generating socialism in the latter. The first has to do with civil rights. In a benevolent environment conventional civil rights are closely related to truth seeking. The truth is often threatening to established positions, even to the positions of great revolutionary heroes. In a society in which the people have the education and the energy and the humane values to understand the truth when it is spoken, that right to speak and disseminate ideas must be given full protection. Conventional civil rights are also closely related to the relations of production. Exploitative surplus extraction may occur even in a basically socialist society, and its victims, whether individuals or groups, must be in a position to make effective protest, or the exploitation will no doubt continue.

No radical would argue that monopoly capitalism represents a benevolent regime. But a well-informed radical is compelled to admit that conventional civil rights are better respected in every monopoly capitalist society than they are in any contemporary socialist society. One understands why this is true—the last section contains part of such an explanation—but conventional civil rights have only been established historically after centuries of struggle, and then where they do occur, only imperfectly. This is a fundamental value that must not be lost in the revolutionary process.

The second difference has to do with nuclear war. The United States essentially has the capacity to blow up the world. It is almost true that one man, the president, has that power. Monopoly capitalism is an irrational society, in which the values, particularly those of the narrow tier of leaders at the very top, are sharply distorted and inhumane. This situation poses something of a doomsday threat of an irrational use of nuclear weapons in a crisis situation. It is a factor that cannot be overlooked in considering alternative ways of generating socialism in the United States.

Fortunately, at least the first of these differences has a positive as well as a negative side, from the point of view of feasible transitions. Radicals obviously do not compete on equal terms with liberals and conservatives in their access to the media of communication in the United States. Nevertheless, they do have fairly substantial access, especially via the university and the printing press. Furthermore, an extended period of respect for civil rights, combined with a substantial level of education in a populace, creates an environment of relative openness to dissident ideas. This environment is not all that un-favorable to the propagation of radicalism. Indeed, one might well argue that the hindrance to further development of the radical movement of the sixties lay more in the relatively unsophisticated quality of radical arguments than in suppression of the ideas. Education and propaganda can play a central role in this environment.

But clearly by itself this is not enough. In the United States radicals are offering a direct challenge to a class that has demonstrated a willingness, to the occasional point of eagerness, to use force against its enemies. The "salami tactics" of a slice at a time have produced dramatic changes in American society: To develop a serious threat a few generations ago of instituting the changes in society that have come since then in a piecemeal way would surely have been to pro-voke a strong repression by the ruling class. The problem is to design at least interim continuation of that approach.

An example will illustrate one possible strategy. The single great-est set of oppressive acts by American monopoly capitalism are those that generate continued misery in the Third World. This has become very widely recognized among American liberals; indeed, they have gen-erally led the piecemeal and ineffective fight against many such prac-tices. But this is an opportunity to separate the process of delegitimiza-tion of policies from the delegitimization of the rule of law, of respect for individual civil rights. The package of policies that produce these sad outcomes are separable from others; they have to do with military aid and training, with politically controlled foreign aid, with policies of support for multinationals, with control of international lending agencies, and the like. A campaign to delegitimize these policies, us-ing an escalating series of instruments, has some chance of success. The initial weapons in such a campaign would be those of research and advocacy, with a view to building a base of support. Instruments would then move to the formal arena of politics, and from there to coordinated civil disobedience and, finally, to direct action against agencies and individuals who continue to administer the delegitimized policies. No doubt at least the credible threat of the latter acts would be necessary to success; probably a good deal more than threat would be required. But once the intolerable costs of this behavior had been properly presented (they often are not effectively presented in radical literature today), this could be a potent and successful action. Fur-thermore, it is in the tradition of revolt against the intolerable, while it continues to respect the institutions of tolerance of dissent in their normal spheres of action. The appeal would be to the undoubted truth of the intolerability of this behavior by American monopoly capitalism. Of course, support from radicals elsewhere would be most helpful; but basically this would be an American job.

Proposals for action outside a very specific political context al-ways have a slightly utopian ring. The above is offered partly in the belief that it *is* both feasible and desirable and partly to point out that there are halfway houses between nonviolence and classical revolutionary violence. If the above program were successful in that it substantially increased the rate of introduction of socialism to the Third World, it would no doubt have substantial impacts on monopoly capitalist countries too, and these could be made to be positive, particularly where dependence on Third World supplies of commodities are crucial to the maintenance of the structure of exploitation in the metropolis. Like all effective revolutionary strategies, it combines research, propaganda, and direct action in the service of a relatively well-defined goal.

From Socialism to Socialism

At the present time there appears to be a growing number of radicals who feel that none of the existing socialist countries are really socialist. No doubt they are correct. A socialist cannot help but be disturbed by the oppressive Soviet bureaucracy and a détente that has some of the properties of a sellout, of the growing inequality and the vulgar commercialism that are by no means minor parts of the Yugoslav scene, of ping-pong diplomacy and the apparent reversal of the cultural revolution in China, of the romantic elitism of the Cuban leadership.

But then not even radical intellectuals expect socialism to be a utopia. A socialist society is still a society with its own laws of motion, its own set of legacies from the past, its own specific set of current problems. A socialist is still a human being with the well-developed capacity to make mistakes inherent in the species. The real question is, In these socialist countries, have substantial and irreversible steps been taken in the direction of real socialism? This is a fundamental question, because no one doubts that the revolutions in these countries were real socialist revolutions, carried out by people committed by lives of desperate struggle and serious study to the promotion of socialism. If such revolutions can succeed politically, but then mostly fail in the aftermath, the case for revolution itself is greatly weakened because the cost of achieving socialism has become almost impossibly high. However, it seems that the correct answer to the question is yes.

In the first place, the elimination of the capitalist class from its power base removes a major threat to the humanizing of human relations. The nature of life in a society that is not ruled by finance capital is very much different from life elsewhere. The so-called new class is a class that everywhere in socialism has seen to it that the primary needs of the people for secure supplies of food, clothing, shelter, and health are met; this too is very much different from life elsewhere. And insofar as long-run trends in these societies can be discerned, they are favorable The time of struggle is not over because these societies have gone socialist, but the terms of the struggle are dramatically changed and seem generally more favorable. In none of these societies had any tradition of democratic politics been developed, and they still do not rank high on a conventional scale of democracy. But the elites seem to be enlarging rapidly, to have a fundamental—and virtually unique in the modern world—concern for the bottom half of the income distribution, and to be opening up politically to the serious discussion of alternative socialist policies. No doubt without further struggle there will not be much more progress. But, unlike the capitalist world, there is a serious case for reform rather than revolution as the optimal strategy in these countries.

Conclusion

Recipes for the transition to socialism are not to be found in this chapter. All that has been attempted is to appraise two or three key aspects of revolution that are of wide, but not of universal, application to the contemporary world. Leninist parties are not a necessity in every part of the world. The mixed strategy will not apply to every monopoly capitalist land; indeed, it may not apply at every point of time in the United States. And some existing socialist countries may well require a revolution to keep them on the track toward genuine socialism. These are issues that must be resolved in socialist debate, which in itself is bound to be a major factor in the further movement of the world toward socialism.

CHAPTER 12

The Future

FOR A LONG TIME, radicals were reluctant to discuss at length the problems of the future society they were hoping to create. It was felt, with good reason, that those who had successfully created the political revolution would be far better able to deal with the morrow of the revolution than were those immersed in the struggle against a capitalism whose demise was still far away. Marx wrote relatively little on the subject, and Lenin's major piece of analysis of socialism was written on the very eve of the October Revolution.[1]

However, the rise of socialism has tended to erode this attitude. There is by now enough historical experience of several socialist societies so that an analyst can develop some feel for the tendencies built into their social structures; and, clearly, there are a number of extant variants of socialism. Attempts to come to terms with this experience cannot fail to have a moral dimension, a concern with whether any given socialist society is headed in the right direction. And that in turn implies some sort of notion as to what the future ought to be like. In addition, the twentieth century is posing a number of problems in new ways, and it becomes relevant to ask whether, for example, socialist societies are well equipped to deal with such issues as ecology, resource exhaustion, the rise of a technocratic elite, and the risks of nuclear war.

There seem to have been three distinct types of utopian societies that have aroused interest.[2] One emphasizes the relation between man and nature as a creative interaction and favors a simple and wholly decentralized society of mutual cooperation in support of this fundamental, "natural" harmony. Essentially, this is a form of anarchy and is epitomized in William Morris's utopian novel, *News from Nowhere*. The second emphasizes relations between humans, solidarity or brotherly love, and finds in the community the primary supporting institution. Many of the communes formed in recent years in Berkeley and Mendocino, California, have been based on this principle. And the third emphasizes man's competitive, even combative, nature and looks to the environment of voluntary exchange as the institutional basis for a realization of this drive. Libertarian anarchists support this utopia. One can find attempts to specify each of these utopias in some detail in the literature; in fact, the student of economics will have already come across arguments in favor of the third of these, for the competitive capitalist regime is the "utopia" on which much of conventional economics (but of course not contemporary economies) is built. Those who prefer one of the other utopias as a general social goal thus have another good reason to be turned off by this apologetic literature in the guise of science.

Of course, no real world society could operate, at least in the foreseeable future, as a pure case of one of these utopias. Every society will have some elements of each of these three fundamental facets of human behavior. But with this qualification the radical position does make a choice among the three. Overwhelmingly, radical literature emphasizes solidarity and community as fundamental bases of human existence; it would be very difficult, I think impossible, to conceive of a radical solution with any other central thrust. From this context we may very briefly survey some issues the future is likely to bring and one or two properties of a radical solution.

The family has been throughout history a central human institution and a primary source of such solidary relations as human society to date has afforded. It will clearly survive into the period of transition to socialism and probably a good deal longer than that; nor is there a reason in principle to oppose the continuation of an institution whose central orientation has been so strongly communal. The basic problem seems to be not with the family but with the institutions in which it has been embedded, and particularly those of monopoly capitalism. In that society, the family has been an institution that has supported deep exploitation of women and children, and has played its role in generating the fragmented personalities that are capitalism's ubiquitous product.

However, historically the family has taken many forms. Among the more interesting have been the traditional extended families, such as the Slavic *mir* and *zadruga*, in which sometimes tens of people representing several generations and many nuclear family clusters have lived together, sharing their property, their work, and their lives. This historical demonstration that groups far larger than the family as we know it today can thrive offers considerable hope for the future. The sharing of work among families united in production units, such as teams and brigades in China, suggests, in an admittedly transitional form, the possibilities for developing social units that generate both solidary human relationships and are economically viable.

There is a crude version of Marxism that holds that the social relations of production are a deterministic product of uncontrollable changes in technology. When this is combined with the observation that the efficient scale of production of many of the most important products of modern civilization is increasing rapidly, the conclusion seems inescapable that before long we will all live under the sway of some giant bureaucracy, for only such a beast could administer the highly centralized economy. The socialist utopian would then have to build his solidary community on a truly grand scale, and serious doubts would inevitably arise as to whether genuine comradeship is truly attainable on such a scale.

However, as we have seen, this notion is rejected by most contemporary Marxists. In fact, there is a growing body of evidence that technology and technical change function as integral parts of the social system, so that changes in the latter can induce changes in the former. Centralization and increasing scale occur because monopoly capitalism finds such a technology to be a useful support for its highly centralized control of the economic system. The movement to socialism will inevitably be a movement toward a technology that is compatible with the socialist system. Once again, in China one can see successful steps being taken to disperse production, even of iron and steel, to the countryside by means of effective small-scale technologies. As countries that have more resources per capita available for the effort turn to socialism, we can expect a substantial acceleration of this process.

Closely related to the nature of technology is the ecology issue. Here the relation between humans and nature is mediated by technology, by our knowledge of how to do things. And it is likely that this interaction will become increasingly unable to be supported by such a mechanism as the price system, in which, for example, pollution costs never seem to find a market price. Can socialism deal successfully with such a fundamental question? Certainly some would say that socialism's success to date in this area has been rather mixed, with the Soviets, for example, having rather similar difficulties in controlling pollution of air and water as the United States.

However, that is a rather misleading perspective. Like other socialist countries, the Soviet Union, starting out as a relatively backward land, economically speaking, has had a lot of catching up to do before many resources could be made available for needs other than the basic material condition of the populace and the economic and military needs of survival. As these additional discretionary resources are now gradually beginning to become available, we will get a better test as to whether it can handle the problem successfully. Even then a failure would not be decisive, given the Soviet status as, at best, a deformed socialist society.[3]

No one can predict the future of a crisis-ridden world with great confidence. But major preconditions for successful action on the ecology front do seem to be built deeply into the basic principles of socialism. Fundamentally, ecology means taking account in one's own behavior of the effect of one's actions on others. Its peculiarity stems from the fact that technology—such as the polluting factory—mediates the relation of cause and effect in this social relation. But in a solidary community, that is precisely where the center of attention is located. An awareness of community is the first and the most important step in resolving any particular ecological problem. Indeed, the requirements of ecological balance will no doubt play a very important role in defining the scope and forms of community in a socialist society.

Manufacturing the capacity to wage a genocidal nuclear war is yet another of the technological triumphs of monopoly capitalist society. In a socialist world one might reasonably expect that physicists and their engineering sidekicks will find something more rewarding to do than to heap one destructive "defense system" on top of another. But the nuclear genie will not go back into its bottle and disappear. One of the most central tasks of socialist states will be to find some way to rid the world of the consequences of our having eaten this particular fruit of the tree of knowledge.

Clearly, that task will not be easy. The uneven development, both economic and in levels of consciousness, of socialist societies will continue to produce tensions well into the era of universal socialism. And obviously some socialist states have nuclear capability. Is there any reason to suppose that they will be willing to dismantle their bombs and rockets, to foreswear their use forever? There are some hints at least that this will happen. No socialist state has as yet actually used the bomb in war. Socialist states have promised that they will not be the first to use nuclear weapons. And disputes among socialist states have so far entailed a use of violence at a far lower level than has been the case under capitalism. But more important than all this is the change in attitude that a move toward socialism entails. The economic security, the level of education, the orientation toward human values, the social structure—all these suggest that tensions will be far more manageable under socialism than under capitalism. The basically defensive foreign policies of the great socialist states is a harbinger of this change, still weakened by the relatively primitive nature of contemporary socialist society, but having its effects nonetheless. Even in this most difficult area there are grounds for optimism.

A socialist world is not inevitable; clearly, it is possible that it will never make its appearance. If it does appear it will only come after prolonged and intense struggles, after a continuation for many years of the crisis we are now all experiencing. But if this analysis has been correct, then it *can* come and it *should* come.

PART II

Commentary

CHAPTER 13

Radical World View and Radical Economics

WHAT MAKES an economic world view tick? Of course, to be viable it must fill some needs of those who believe it. I suppose the most natural way for an economic world view to fill a need is to provide a correct interpretation of how contemporary political economies work. But as anyone who knows the elements of Marxism is aware, there is not a straightforward relationship between subject and object, between the material world and our awareness of it. That relationship is mediated by our experiences and by our social conditioning, as well as by hard facts. And no one, which means no Marxist, no radical as well, has as yet escaped some measure of social conditioning and some measure of exposure to unique experiences. This suggests that at any particular time the world will contain several economic world views, each of which will seem most plausible to some.

However, the collection of plausible radical economic world views has one essential advantage over any competitors, an advantage that stems from the very nature of the radical world view. As was noted in the early chapters of Part I, radical interpretations are built on an appraisal of the fundamental structure of society, and in particular on the class nature of society and its implications. By going to the root of the matter right from the first, the radical interpretations are able to avoid a good deal of the obfuscation and distortion that are entailed by attempts to paper over this most fundamental of social facts. This is the distinguishing feature of radical economic world views, and also constitutes the major advantage they possess vis-à-vis rival interpretations. As a consequence, in Part II we will not be much concerned with these nonradical interpretations.

Such a limitation will not free us from the need to appraise differences for, as the reader of Part I is certainly aware, there are other plausible radical economic world views than the one presented there.

In fact, there are several quite well-developed radical economic world views, and they tend to differ fairly substantially from one another. Essentially, Part II is devoted to looking at various alternatives to Part I and explaining our choices. Before beginning the job, however, something must be said about the process of choosing among alternative world views.

There are probably three criteria that are central to this choice. The first test of the alternatives is simply how well each one fits in terms of one's own experience of life and interpretation of history. I suspect that this criterion is mostly applied intuitively, but there is no reason why one could not make a systematic appraisal. The second criterion is a bit more precise: The question is, Do the world view's assertions fit the known facts? In practice, this criterion turns out to be rather less specific than it sounds because of the nature of the "facts" that appear as parts of so broad a doctrine as a world view. For example, one of the most striking facts asserted by Paul Baran is that in Third World countries the relative surplus is quite large. To my knowledge this thesis has never been seriously tested. The concept of surplus is itself underdeveloped, and global statistical investigations have so far been carried out almost exclusively by liberals using different concepts. Nonetheless, as was argued in chapter 8, it is a very plausible fact, and certainly has not been refuted.[1] It would be quite irrational to reject a world view because it contained this assertion as an integral part of the package.

The third test is potentially the most useful. A radical has already narrowed his range of choice to a segment of the available world views and has eliminated those that patently contradict experience. The remaining test has to do with whether or not the various parts of the world view fit together to make a coherent whole. This is not a simple criterion to apply because there is a certain open-endedness to it; for example, a revealed incompatibility may be relatively minor, or one might feel that it can probably be corrected without too much difficulty. But major incompatibilities are serious obstacles to the acceptability of a world view.

An example will illustrate the issue. Baran and Sweezy have argued that under monopoly capitalism there is a strong underlying tendency toward economic stagnation, combined with a need for the economy to expand to prevent increasing unemployment from becoming a threat to the system. This leads in turn to much waste and to the development of a fundamental problem, namely, the problem of absorbing the tremendous surplus the system is generating.

The arguments they present in favor of this thesis are in themselves plausible. However, there is one implication that is very hard for a radical to swallow: The thesis implies that there exists a reformist solution to the problem of imperialism. That is, if surplus absorption is the problem domestically, it can be resolved, for a long time to come, by a massive foreign-aid program that diverts the surplus to Third World countries. Domestic capitalists could, in effect, be bribed with guaranteed profits to develop the Third World as a response to a basic domestic contradiction of monopoly capitalism.

This reformist thesis is inherently implausible as a characterization of imperialism over, say, the last half century. And it would seem that the incompatibility is fundamental: Either buy that or reject the thesis about the wastefulness of monopoly capitalism. But the alternatives are not in fact that sharp. As will be argued in chapter 15 below, by rejecting the idea of underlying stagnation in its Baran/Sweezy form the plausible parts of their thesis can be retained without being forcibly tied to a reformist theory of imperialism.

In the world view appraisals to follow, the compatibility test will play a very important role. Unfortunately, there has been no systematic effort to appraise radical world views seriously and with a view to developing optimal versions. Though one finds book reviews and the like, these are mostly unsystematic impressions and are usually not devoted to an appraisal that is concerned with world view development. As a consequence, our efforts are inevitably selective and, I'm afraid, somewhat introductory. It turns out that world view appraisal is a fairly serious and demanding business.

What is the relation of a world view to the scientific part of economics? The notion of science is nowadays fairly well defined. It consists at the very least of theory and observation in fairly close interconnection, with serious attention devoted to the establishing of the facts as a central part of the discipline. Clearly, world views by themselves cannot pass that test. Nevertheless, as will be argued, they do have an important relationship to the scientific part of a discipline.

Scientific radical economics has been under intense attack throughout the twentieth century. The political constraints on its practice in most socialist countries are substantial. In some capitalist countries there is widespread suppression of radicals and their work. In areas where there is toleration, there is a serious shortage of resources for research, especially when compared with the resources available for those scientists whose work essentially defends the system. And radical scholarship is further decimated by the felt need of many potential scholars to devote themselves substantially to political work.

But even under all these handicaps, scientific radical economics is definitely on a strong upward trend these days. One of its major tasks is the reinterpretation of the massive body of essentially liberal work so as to reveal the true meaning of that research. And it is at just this point that a world view becomes practically useful. A good world view provides the researcher with an orientation specific enough to guide his work of reinterpretation, suggesting hypotheses and approaches and, perhaps most important, providing a firm specification as to what *is* radical to sustain her in his struggle with essentially alien materials. Much the same also holds for new, as opposed to reinterpretive, research.

The role that a world view can play in determining one's approach to problems can perhaps best be illustrated negatively, that is, by looking at a serious and intelligent appraisal of "The Political Economy of the New Left" in a very popular book of that title by the Swedish economist Assar Lindbeck. A social democrat politically and a thoroughgoing liberal in his economics, Lindbeck obviously did his homework, and even seems to have tried hard to be sympathetic in his appraisal of latter-day radical economics. Nevertheless, his book leaves the reader with the impression that the New Left economists are essentially a collection of well-meaning bumblers, of people who don't do too bad a job of asking questions, except that they never bother to answer them seriously.

This conclusion is a very straightforward consequence of the method Lindbeck uses. And the method is very liberal. It consists of dividing all questions into relatively small parts, the parts being recognizable as questions typically asked by liberals. For example, Lindbeck finds radical discussions of exploitation of Third World countries unsatisfying.[2] She points out that radicals emphasize the harmful effects of high profits made by multinationals in developing countries, when in fact high profits are merely a sign that the company is efficiently providing a desired service. Lindbeck has thus implicitly reposed the problem: He claims the issue is, Which is better, an efficient or an inefficient multinational? I suppose most radicals would not be too averse to accepting Lindbeck's answer to *her* question. But they also would not find it to be at all an interesting question. The reader of Baran, or of Part I above, will know that the efficiency of the multinationals is not a central issue when it comes to center-periphery relations. The question there has to do with the existence of a pattern of exploitation, based on a large number of different instruments, many of them political and military, many of them economic, many of them of considerable age, many of them involving deep penetration into the social structure of the Third World country. By asking the questions from within the framework of a liberal rather than a radical world view, Lindbeck both distorts the facts of exploitation and defuses the outrage that those facts tend to generate in the reader.

The main thrust of radicalism is to bring about fundamental change in an exploitative society. This goal requires that one ask different questions from those asked by liberals, and the questions in turn often require the use of different techniques in order to answer them. Liberals want to preserve society, dealing with emergent problems by making small changes that will not disturb the basic structure. Those are very different goals, and they produce a very different structure for the interpretive science of society that supports each effort.

However, there is more to the difference than the nature of the questions. These two world views are oriented toward the needs, experiences, and values of two different classes. Out of that more fundamental difference lies the gap that separates members of these two classes from one another, making communication across the gap very difficult. Since the differences are based both on life experiences and on very powerful socialization processes, they may never be fully bridged. That failure, and all it entails, is simply a cost of capitalism —but not of socialism.

A developed and integrated world view can play an important role in formulating and spreading understanding of the nature of the world among its citizenry. And that in turn is the basis upon which the elimination of exploitative systems can taken place. However, at this point a problem emerges from within the radical movement. It seems that many radicals believe that serious attention to the intellectual side of radicalism is largely a waste of time, even a cop-out: "The basic ideas are already in Marx and Lenin; what is needed is to abolish capitalism, not study it." I suppose that if you expect the revolution to be coming next year at the latest, that is a not implausible view. And in fact, the way things are going there may very well be a successful political revolution somewhere while this book is in press. But that misses the point on two counts. In the first place, there is no basis for feeling that the revolution will be coming in the heartland of monopoly capitalism in the near future; in the second place, there will still be a central need for both radical world views and scientific economics after the political revolution. A central function of radical economics is as an instrument of recruitment. But an equally central function is to provide a correct understanding of how the world works. Without that, there is no good reason to feel that radicals will be able to bring about a genine revolution. After all, it is no mean trick to pull one off. And this applies equally after the revolution. Revolutions can go astray; the revolution is itself both an end and a beginning. If it does go astray, or begins to, there must be those who can point this out, and they must have the same sort of basis for making their case as their prerevolutionary forbears did. One has only to look at the Soviet Union, where such efforts are continually and brutally suppressed, to see the nature of the need. The construction and development and appraisal of radical world views is more than intellectual game playing. It is a serious and vital part of the overall radical effort.

CHAPTER 14

Baran and Sweezy

I THINK it is not controversial to assert that the two Pauls, Baran and Sweezy, have been the most substantial and most influential radical economists writing in the United States since World War II. Baran's book, *The Political Economy of Growth*,[1] is probably the most important single work of radical economics of the postwar period. It contains an integrated characterization of the functioning of monopoly capitalism, of the functioning of capitalism in the Third World, and of the interaction between the two. The joint work of Baran and Sweezy, *Monopoly Capital*,[2] extends and to some extent revises the account of monopoly capitalism, and has also been very influential. Both authors have, of course, extended and refined more detailed aspects of their views in articles, some of which have been collected and published in book form. Sweezy, as the founder and continuing editor of the Monthly Review Press, has also provided an outlet for a large fraction of the most important radical works of the postwar era.

Because of the similarities in their viewpoints and their collaboration, it is convenient to consider these two authors together. In combination they have a coherent and broad-gauge perspective, one that can reasonably be considered to be the appropriate starting point for anyone trying to come to terms with postwar radical economics. However, that is not to say that every radical economist will find all of even their major arguments acceptable. Our evaluative procedure will consist of the following: The next section outlines the views in Baran's *Political Economy of Growth*. Then the extensions and modifications of those views as they appear in *Monopoly Capital* and elsewhere are described. The final section of this chapter indicates the major points of agreement and disagreement between Baran and Sweezy and the optimal radical economic world view. The two following chapters appraise further key aspects of their work, including some comparisons with other radical writers. In reading the next section, the reader should remember that the illustrative examples were selected by Baran around 1955. They have been included here because they in no way detract from the modernity of this prescient work.

Political Economy of Growth

The basic concept Baran uses in analyzing contemporary societies is potential surplus. This is the difference between the output that could be produced, if all currently available resources were put to use, and essential consumption. At present it is not possible to make good estimates of the magnitude of the potential surplus, though a number of important qualitative statements can be made. But first a further definition: Three things can be done with the surplus—it can be invested, consumed, or wasted. The way in which a society basically functions is revealed by the ways in which it allocates surplus among these three categories, while the size of the surplus is largely a matter of the forces of production.[3]

Nineteenth-century capitalism became a rapidly growing economic system, Baran argues, because it was able to meet four basic conditions: (1) unemployment of resources was kept to a minimum; (2) the wage rate was kept down to a level that tended to maximize the size of the surplus; (3) a maximal share of the actual surplus was plowed back into investment rather than being consumed; and (4) there were plenty of profitable investment opportunities. The effect of the first three conditions tended to counter the effect of the fourth in the capital market, thus keeping the interest rate relatively low.[4]

With the rise of monopoly capitalism, including especially a high concentration of ownership and control in the modern industrial sector, this situation changed; in fact, none of the four conditions continued to hold. The volume of unutilized resources became very high, the organized labor force was able effectively to bid up wages, capitalist and other forms of consumption increased at the expense of further investment, and the number of investment projects that were deemed worthy of realization by capitalists declined sharply. Most of this change is attributable to the change in the nature of the capitalist system itself. Thus it is not true that in a rationally ordered society there would be relatively few investment projects these days; rather, it is a matter of monopoly capitalists viewing the profitability in a different light. For example, they are rarely willing to undertake an investment project that will destroy the value of some of their existing capital stock, whereas in the older, more competitive regime, the capitalists would be forced to adopt such a project in order to survive.

Thus there is a strong underlying tendency toward stagnation in monopoly capitalism. This creates serious problems. For example, with productivity and labor force both growing at a couple of percentage points each year, either some means of using the additional hands must be found or unemployment levels may begin to create a revolutionary situation. In fact, the rise of welfare state spending tends only to occur during such crisis situations, and partly accounts for the big state under monopoly capitalism. Another reason is the finally recognized need to regulate the economy in order to try to prevent the survival threat posed by major depressions. Already by the end of Roosevelt's administration it was clear that business was taking over the administration of this new state function.

The newly emerged state system with its substantial welfare spending and the great waste of resources that nevertheless keeps unemployment rates at a manageable level, combined with imperialist operations abroad that accrue to the national benefit, creates a kind of people's imperialism, "a far-reaching harmony between the interests of monopolistic business on one side and those of the underlying population on the other." [5]

However, the stability of this system is highly precarious. The increasing waste in the form of excess consumption of gewgaws, unproductive investment in these goods, the excessive costs of organization of monopoly business, government programs of military expenditures and the like, creates an irrational society that can hardly escape the notice of the relatively deprived citizenry. Clearly, much of this kind of expenditure cannot be increased indefinitely; but increase it must if the steadily growing surplus is to be absorbed. Effective control of the economy has not yet been achieved, with inflation posing a special problem because

by causing the development of a cleavage between the interests of creditors and debtors, by dispossessing the new middle class and the rentiers, by depressing the real income of the workers, it seriously weakens the authority of the government and disrupts the political and social cohesion of the capitalist order. . . . Thus the stability of monopoly capitalism is highly precarious.[6]

The tendency toward caution and circumspection that monopoly capitalists exercise in their business affairs seems to have spilled over into their international behavior, in that they seem to prefer cold wars to hot wars; however, this line of argument cannot be pushed too far because of other structural changes in the economy. For the first time in its history the United States is now getting a " 'full-time national-scale arms industry, [so that] companies . . . now treat their war output as a permanent part of their business,' " as *Business Week* put it.[7] This tends to push the government toward exerting its muscle, so as to justify the increasing arms expenditures. And the forces pushing toward war are profoundly irrational, not subject to easy control "as many big business magnates in Germany discovered to their sorrow." [8] It is this profoundly irrational system that monopoly capitalism has unleashed upon the world.

But the world had already been "prepared" for this onslaught by previous events. One of the most striking was the creation of underdevelopment. In much of the world, including precapitalist Europe, three basic preconditions for the emergence of capitalism were being created. These were: (1) a slow increase in agricultural output combined with intense feudal pressure on a servile (in status) population, and displacement and consequent rebellion of peasants; (2) the continuing process of division of labor, together with the emergence of a potential industrial labor force, the evolution of a class of merchants and artisans, and the growth of towns; and (3) a fairly spectacular accumulation of capital among merchants and wealthy peasants. And in the late seventeenth and early eighteenth centuries contacts between Europeans and non-Europeans seemed to foster the spread of science and modern technology.

But except in a few special cases, such as the United States and Japan, capitalism and the industrial revolution did not emerge in the rest of the world. Why not? In part because the military superiority of Europe permitted the extraction of resources from these countries, which served as a large fraction of the primary accumulation that fueled the European expansion while impoverishing the rest of the world. In part because the intrusion of Europe disrupted the internal dynamic of these societies, so that the social forces moving toward modernism were unable to effect their transformations. The United States, like the Commonwealth countries, escaped this process because they were, in their controlling populations, primarily European. Japan escaped because of the combination of her own earlier isolation, which protected her from the worst of the depredations and left her social structure intact, and because of the preoccupation of imperialist powers elsewhere during the key period of the Meiji Restoration. There was no escape elsewhere in the world, and the rest of the world remains what it was made by early predatory capitalism: underdeveloped.

The material conditions for rapid economic development exist in the underdeveloped world today. The first two conditions (page 275) of growth, large relative surplus and low wage rates, both apply, the first even in the countries we think of as being poor. The problem is with the latter two conditions. To see this one must first look at the class structure in these countries. The populations are primarily engaged in agriculture, where a relative surplus of perhaps one half of the output of the sector is generated. However, this surplus is appropriated essentially by four groups: (1) merchants, moneylenders, and other intermediaries; (2) local capitalists controlling industrial production; (3) foreign enterprise; and (4) the state. The first category is clearly parasitic. The second usually involves small morsels of capital and so does not use the best available modern technology, tends toward monopoly, and often is not growth-oriented. The third produces the familiar enclave, oriented toward the world market rather than domestic needs, with its managers skilled in the arts of tax avoidance and other forms of corruption. They engage in substantial political interventions aimed at generating a favorable climate for investment, which often means building up substantial armies to protect the foreigner against the pauperized masses of the country.[9]

There are three kinds of state regimes, colonial, comprador, and New Deal. The failures of colonial regimes to generate economic development is obvious enough. Typical of the compradors are the Middle East and Venezuelan oil kingdoms. They earn vast royalties on the oil extracted by foreign companies but have been spending it, for the most part, on almost anything but economic development. Much of the income slips back to the monopoly capitalist homelands one way or another. New Deal regimes are under the control of a nationalist bourgeoisie with various fringe allies. They can be growth-oriented and under favorable conditions may be able to reproduce the Japanese escape from underdevelopment. But the dice are heavily loaded against them by monopoly capital control of world markets and modern technology, and the outlook is not favorable. Their basic problem is that the high rates of investment needed to modernize can only be achieved by government mobilization of a portion of the relative surplus; but this requires heavy taxation of the regime's own principal supporters, not a popular policy among politicians.

This analysis of underdevelopment produces three conclusions that conflict with typical liberal arguments: (1) the development problem is not primarily that of a shortage of capital but that of generating investment, instead of consumption and waste, out of the large relative surplus; (2) there is no shortage of entrepreneurial talent in the developing countries, but rather there is an absence of a structure within which entrepreneurship can function; and (3) the population problem is not basically one of finding ways to control population growth, but rather of creating a rational society within which the participants will be encouraged to make rational choices about their lives.

The truth of these propositions can be examined negatively by looking at the failure of capitalist development processes in the twentieth century; it can also be examined more positively by looking at the way in which development is achieved under socialism. "The socialist camp, preoccupied with internal construction, is utterly unlikely to initiate a war." [10] The waste of resources aimed at propping up reactionary regimes in the socialist countries is converted into investment, and the high investment rates produce high rates of growth. Much of this surplus, of course, comes from the agricultural sector, where agricultural production is transformed into "specializing, labor-dividing and market-oriented industry" in which planners can control the process of surplus extraction in a rational way. The extensive policy of concentrating investment initially in industry rather than agriculture is most productive in the long run. There is still some waste in these regimes, but it stems largely from the continued threat of monopoly capitalism and the resulting deformities in military expenditures and excessive autarky to prevent use of economic leverage against socialism. Once a country has been removed from the grip of exploitation by national and foreign capital, the possibility of successful economic and social transformation in the twentieth century is no longer problematic, as the experience of these socialist countries shows.

In brief outline, this is the essence of Baran's argument.

Modifications and Extensions

The Political Economy of Growth was completed in 1955, more than two decades ago. The years following its publication produced a great deal more information about economic development, about the socialist countries, and about the functioning of monopoly capitalism in the relatively new Cold War environment. The basic ideas of Baran's book survived, and are essentially repeated in *Monopoly Capital,* which appeared a decade later. But some new ideas and some shifts of emphasis as well as some modifications of earlier arguments made their appearance in the later work. We will simply run through a short list of the most important of these.

1.   In his 1962 foreword to the second edition of *Political Economy of Growth,* Baran puts considerable weight on the forces of production as hampering the domestic development of socialist practice in socialist countries. The political troubles of socialism he attributes to the slow rise in consumption necessitated by the need for a rapid growth of output. Resistance to this "creates the need for political repression" and for the tension between socialism and democracy.[11] The Soviet thaw of the fifties he attributes to the further growth of the economy. Similarly, he finds uneven development to be the key to the problems in relations between socialist countries. The China-Russia tensions are due to the fact that China is not yet economically ready for the thaw. Once again, further growth of the productive forces will mitigate these anomalies in socialist behavior.

2.   In *Monopoly Capital* Baran and Sweezy complement the earlier book with respect to the impact of the managerial revolution. They come down strongly on the side that perceives this evolution as primarily a matter of administration and not at all a matter of the rise of a new class. The managers are agents of the ruling class, and their behavior reflects a continuing primary interest in generating profits. This is quite consistent with Baran's earlier position, but in the later context represents a firm denial of the validity of the "finance capital" thesis that central financial institutions have come to control large American corporations.[12]

3.   There is more emphasis in *Monopoly Capital* on stagnation than in the earlier work. "The *normal* state of the monopoly capitalist economy is stagnation." This is accompanied by downplaying the Steindl thesis that had attracted Baran. Steindl was an early proponent of the idea that the rate of innovation is closely tied to investment rates, since the innovations can only be realized in new capital investments. He added to this thesis the inverse causation, namely, that low rates of investment tend to inhibit the innovative process because of the absence of expected outlets, while high rates of investment stimulate innovation. Baran and Sweezy now emphasize that there is no necessary connection between the rate of technical progress and the volume of investment outlets.[13]

4. Baran's strongest argument for the wastefulness of capitalism was the indirect one that, judging from the situation during World War II, the nation seemed to have from half to three-fourths of its output more or less disappear without reducing real wages.[14] In the later book the two authors develop the direct argument further, that is, by pointing to the various ways in which waste occurs within the production sector. Perhaps their most interesting story along this line comes from a study that estimated that the per-car cost of essentially cosmetic model changes in the automobile industry over seven or eight years amounted to more than forty times the advertising expenditures per car.[15]

5. In the later book much more emphasis is placed on the limits to wasteful surplus absorption under monopoly capitalism. Basically, the argument is that military-imperialist spending has had to bear the major burden of surplus absorption, and that the ability of this sector to continue to increase more rapidly than the growth of output is itself limited. Civilian government was felt to have already reached its limits as a relative surplus absorber because of the conflicts among powerful interests created by civilian spending, which tends to displace private enterprise as military spending does not. The sales effort can continue to be an absorber, but not without simultaneously generating increasing awareness in the population of the extreme wastefulness of this process. Investment cannot be relied on as a relative surplus absorber, except when a truly fundamental innovation comes along, such as the railroad. And capitalist consumption has tended not to increase relative to the growth of output. This suggests strongly that a crisis of irrationality is brewing, though it does not permit any great precision in defining its nature and timing.

6. Discussion of imperialism is extremely brief and emphasizes the military spending side. It is pointed out that successful foreign investment tends to increase the surplus that must be absorbed. However, it is in the interest of the individual corporation to make these investments, thus producing a contradiction between the parts and the whole that is of the essence of monopoly capitalism.

7. In discussing the history of monopoly capitalism, Baran and Sweezy emphasize the tendency toward stagnation as arising early; it would have been apparent by the 1880s were it not for the railroads, which absorbed almost half of American investment for two or more decades. They also emphasize the role of great catastrophes in masking the nature of monopoly capitalism during the twentieth century, and particularly the two great wars and the Cold War as concealing for extended periods the stagnationist tendency.

8. Race relations and alienation, two areas that played little role in Baran's previous work, are given strong emphasis in *Monopoly Capital*. The divide-and-conquer tactics that work to keep wages down and to prevent strong political action by buying off the black bourgeoisie represent one durable aspect of capitalist relations of production. Another argument runs from alienation to prejudice, the pressures of life in monopoly capitalism tending to find in prejudice a rather

natural outlet. But alienation affects all aspects of life, through the lack of commitment, of involvement, that is inevitable in so profoundly irrational a society.

9. Baran saw a capitalist crisis developing, but he also saw mechanisms at work that could long delay it. Baran and Sweezy do not seem much more optimistic about the possibilities of their crisis coming soon or generating an internally inspired transition to socialism. "As the world revolution spreads and as the socialist countries show by their example that it is possible to use man's mastery over the forces of nature to build a rational society satisfying the human needs of human beings, more and more Americans are bound to question the necessity of what they now take for granted." [16]

10. Sweezy's own position regarding socialism has undergone some modification in recent years. Partly as a result of an exchange with the French Marxist Charles Bettelheim, Sweezy agrees that during the long period of transition to socialism after the political victory of the revolution, markets are likely to play a large and positive role as a means of resource allocation. [17] He had been much more suspicious of markets in the past, regarding them as capitalism's Trojan horse. Also his attitude toward the centrality of participation and democratization at an early stage in the transition to prevent the emergence of merely a new form of class society has seemed to change. This is reflected in some of his critical remarks on one-man rule in Cuba, made in a 1969 appraisal of that revolution. [18]

11. As Sweezy himself pointed out in an article, the New Left may be new, but the people who provided it with its basic intellectual orientations toward contemporary society were the great men of the Old Left. This is particularly true of the United States. But with the shifts of emphasis suggested above, and particularly of those that relate to socialism, Sweezy is himself adjusting in the direction of New Left orientations. As a result this dichotomy has substantially disappeared.

12. An important extension of the Baran/Sweezy view of monopoly capital also requires some comment. Harry Braverman's *Labor and Monopoly Capital, The Degradation of Work in the Twentieth Century*,[19] is an attempt to add a characterization of changes in the labor process to the theory of monopoly capital. The degradation of work constitutes the heart of this thesis. Scientific management and its related schools of personnel administration symbolize this change most dramatically. The worker is now viewed as an object, "a general purpose machine operated by management," as Braverman describes it.[20] The worker's tasks are analyzed with a view toward designing efficiency into their execution in precisely the same way as is done for machines. The function of the industrial psychologist is to design an environment that is maximally supportive of maximum worker efficiency; that is, the worker is once again treated as an object subject to control in the service of profits. These interventions, profound in their human consequences, are also profoundly alienating. But as Braverman's research brilliantly reveals, there has been another profound consequence. The

levels of skill required to perform the broadest range of both blue- and white-collar labor has steadily declined. Braverman exposes clearly the fallacies of statistical definition that have produced the opposite impression in most observers, and enriches that discovery with a mass of detailed accounts of specific changes in the nature of work over the twentieth century. His conclusions are lent added support by the fact that Braverman is himself a worker, having spent about half his work-life in skilled blue-collar jobs and the other half in various skilled white-collar jobs.

Our Part I characterization of the optimal radical economic world view made use of Braverman's thesis, though perhaps did not give it sufficient emphasis. One problem with his work is that some of the key assertions will require further scholarly investigation before they can be accepted without question. This applies particularly to the thesis that on the whole work is more degraded now than it was, say, at the turn of the century in the advanced capitalist countries. A second problem is that Braverman essentially accepts the Marcusan thesis that there must be a dramatic change in the structure of technology itself before anything like socialism can be feasible. This thesis is discussed in chapter 16 below and, I believe on present evidence, is too extreme. But that is in no way to disparage Braverman's work. Her most impressive achievement is that, by posing a well-thought-out and carefully investigated alternative, he has revealed the poverty of a whole generation of conventional research into "labor economics."

Baran and Sweezy and the Optimal Radical Economic World View

Clearly, the world view articulated in Part I above owes a great deal to the work of Baran and Sweezy. The notion of economic surplus, which was first applied by Baran to the analysis of monopoly capital, also plays a central role in Part I, as does the analysis of waste and inefficiency, the idea that a new form of capitalism emerged around the turn of the century; and so too the Baranian analysis of underdevelopment. No analysis of monopoly capital can be complete without their contributions. Of course, some of these ideas have been floating around for a long time, but an integrated analysis, based on a clear view of the distinctiveness of this system, as well as of its similarities with classical capitalism, is the product of their intellectual labors. It is probably the most substantial intellectual achievement by radical economists since World War II.

Nevertheless we have not followed their lead in every respect. At several points, some of them important, we have taken a different tack. The most important of these, the relationship between stagnation and surplus absorption and their joint implications, was mentioned in the previous chapter and will be the topic under discussion in the next chapter. Other points of difference will be discussed later. Our criticisms are intended to be constructive, aimed at developing an optimal radical economic world view. The success of that venture depends on the development of an integrated social process of radical criticism. Hopefully, the time is past when a major work of radical economics can go essentially uncriticized in the journals for years. A world view has important dimensions other than the intellectual. But the intellectual dimension is of central importance, and that dimension must be based on a serious and critical, if basically sympathetic, appraisal of the products of radical intellectuals.

CHAPTER 15

Stagnation and Surplus Absorption

A MAJOR DIFFICULTY in developing an effective interpretation of the operation of monopoly capitalism has been the lurching described in chapter 4. Monopoly capital has proved to be an effective device for the generation of catastrophes. This in turn has tended to screen from the student's eye the basic processes at work within the system. For example, in the nineteen thirties and forties there was a general tendency to emphasize stagnation as an unavoidable property of monopoly capitalism. Not only was there the depression of the thirties, but the last quarter of the nineteenth century had come to be called the Great Depression, at least until a much greater one came along in the nineteen thirties. There were arguments that new investment opportunities were tending to dry up and that the decline in population growth had a depressing effect on capitalism, as a result of the smaller demand of the smaller population. One interesting thesis suggested that stagnation tended to reinforce itself, or perhaps even be created by a period of depression. The idea was that new inventions tended to be stimulated by the prospect of being put to profitable use. But putting them to use required new capital. If investment flagged so would the zeal of inventors.[1]

The stagnationist thesis was not implausible in those days. But from today's perspective it has lost most of that plausibility. Clearly, capitalist economies go up as well as down, and they do the former much more often than they do the latter. There have been several fairly sustained periods of growth since World War I, including the nineteen twenties, fifties, and sixties. Furthermore, if one compares the available statistics on the nineteenth-century performance of advanced capitalist countries with performance during the twentieth century, the growth rates do not support the thesis of flagging growth. Nor do the investment ratios, the proportions of national output plowed back into the building of new plant, suggest a flagging of investment; these ratios have tended to be higher in recent decades than they were in comparable decades of the last century. Nor does the history of technology suggest some flagging of underlying opportunities. Not only is the information revolution in full swing, with clearly very dramatic technological changes in the offing, but the nuclear power industry is probably still in its infancy with the long-term prospect for fusion power promising. And at lower levels in the invention hierarchy there has never been as much activity as at present.[2]

Paul Sweezy continues to subscribe to the stagnationist thesis, as did Paul Baran in his major works. It is still a popular view among radical economists of a variety of persuasions, including, for example, Ernest Mandel. But it seems an unnecessarily restrictive view of monopoly capitalism, and one that clearly has not been established as a product of careful and detailed empirical work. Given that fact and the fact that the crude evidence does not support the thesis, I think "volatility" is a better characterization of monopoly capitalism's movement through time.[3]

There is no need to repeat here the arguments of chapter 9 in defense of this assertion. However, there is some need to relate this difference in interpretation to the current problems of stagflation, ecology, and resource constraints. In the first place, the effect of these three phenomena is to reduce the potential growth rate of monopoly capitalism. The stagnationist prediction of declining growth rates may very well be vindicated in coming years. However, this problem is occurring in a very different context from that of the earlier discussion. Capitalism has lurched into a new state of crisis. This time the resource constraints reflect not a current exhaustion of resources but a conflict between monopoly capital and the Third World, one in which the ability of the Third World to defend its interests and its national domain is increasing. The ecological crisis is a product of the distortions that occur in a regime where production is for profit rather than for use. That has, of course, always been true, but it becomes a genuine crisis only when capitalism has achieved tremendous growth, so that we are literally drowning in the by-products of profit-seeking production activity. Stagflation is one of those turbulences in our irrational system that is not well understood. But clearly it is in considerable measure a consequence of irrational policies of the past, which generate premature bottlenecks, saddle the economy with a tremendously demand-distorting overhang of debt, and divert millions of potentially productive human-years of labor into the system's massive conflict-resolution effort.

Perhaps stagnation is not so bad a term for this, after all. So long as one recognizes that this stagnation has different causes and operates in a different environment than was asserted for the earlier period, and also recognizes that this stagnation is accompanied by increasing volatility in the levels of economic activity, the term certainly retains descriptive power.[4]

However, there is a more serious consequence of the earlier stagnationist thesis when it is combined with the Baran and Sweezy surplus absorption thesis. These two ideas in combination seem to produce a serious distortion of the functioning of monopoly capital. The first, and probably most important, has to do with the nature of imperialism. Let us for the moment accept both stagnation and the Baran/Sweezy version of surplus absorption as the major problem facing monopoly capitalism. And let us think of ourselves as the political leadership of an advanced monopoly capitalist country, considering ways to absorb enough surplus to keep revolution at bay at home, while the system continues to grind out its massive quantities of mostly unsatisfying products. Our problem is only made more difficult by the operations of the multinationals in the Third World, where their ripoff operations are generating and repatriating still more surplus to absorb. What to do?

The reader probably sees the capitalists' presumptive answer. Let the Third World become the sink down which the "excess" surplus flows. A massive program of loans for investment in modern industry and infrastructure, possibly even accompanied by massive grants could, in this context, be of great benefit to the monopolists at home, since they would be expanding to meet the new demand. The jobs created would keep the industrial reserve army at home at its "optimal" or productivity maximizing level, and some small portion of the increased output could be assigned for small increases in real income to domestic workers. For a generation the Third World could be kept in a state of expansion and so perhaps accepting of oppressive, if growth-oriented, political controls. The problem of surplus absorption would thus be substantially mitigated by eliminating the problem of stagnation.

A program of this kind has actually been advocated by some American leaders.[5] However, it has never been implemented. Our foreign aid and loan operations have never been on a scale that could have a significant effect in proportion to the level of relative surplus; also, they have never been increased at a rate that amounted to any significant proportion of the increase in surplus in the United States. If this obviously reformist scheme has been the way out for American monopoly capitalism, it has never been possible to convince any substantial section of the ruling class of that fact.[6]

I think that it is not the ineptness of the ruling class that is responsible. Rather, the thesis represents a misinterpretation of the situation monopoly capitalism has found itself in during the past quarter of a century. Imperialism is a problem central to the structure of monopoly capitalism; it is not something that the capitalist ruling class would be able to eliminate or substantially mitigate if only they would act rationally to serve their own interests as a class. The heart of imperialism lies in the class nature of *both* societies, center and periphery, and in the fact that it is in the interests of the center's ruling class to preserve the national bourgeoisie in power in the periphery. But far from having difficulty absorbing surplus, the various members of the center's ruling class are engaged in a keen struggle among themselves for a larger slice of the surplus pie. The nature of the society, and particularly the "bureaucratic interface" problem discussed in chapter 7, leaves no room for such foreign-aid schemes as the above. In fact, there is keen competition among the monopoly capitalists for an increasing share of the surplus that is still obtainable from the Third World. Stagnation just does not fit this picture.

Another difficulty is raised by the surplus absorption thesis in its Baran/Sweezy form. This is best suggested by the prediction in *Monopoly Capital* that went astray. Using data down through 1963, the authors stated that civilian government expenditures had already reached their peak capacity as surplus absorbers. However, the argument supporting this prediction was rather vaguely drawn and actually did not support so specific an assertion. And the facts almost immediately began to contradict the prediction. In the decade following 1963, civilian government expenditures became the leading absorber of the increasing surplus, far outstripping military expenditures in both average rates and amounts of growth. And, of course, that trend has continued.[7]

What went wrong here? As already noted, the analysis really did not support the prediction. In general, the various assertions in the book about upper limits to government spending, both military and civilian, are unpersuasive. No serious basis was laid for establishing such upper limits. However, it is very important to note that the basic idea of the existence of an upper limit to the share of government spending in total spending under monopoly capital is a sound one. As government gets relatively bigger, the time will come when the system of private property relations will begin seriously to be threatened. At that point a "negation of the negation" process sets in, in which government begins to erode its own effectiveness in carrying out its primary function under the capitalist system. That will clearly be a time of serious crisis. Substantial further relative growth of government spending beyond that level would mean an end to the system of monopoly capital. Clearly Baran and Sweezy were moving in the right direction in their analysis.

These criticisms of Baran and Sweezy obviously need further appraisal. However, that will require serious and detailed study, and cannot be undertaken here. But one point should remain clearly in view. It *is* possible to criticize even fairly fundamental aspects of widely held radical views without losing one's radical perspective. That is the meaning of "constructive criticism" within the framework of radical economics, and a good deal of it is badly needed in order to develop further the plausibility of this system of belief.

CHAPTER 16

Alienation

THERE IS a profound ambiguity inherent in the basic concepts of conventional economics. Consider a term like consumption. On the one hand, it makes a clear and positive reference to relatively easily observable things and events, such as a car or a haircut. On the other hand, it has an essentially personal and subjective meaning, referring to the generation of satisfaction in the individual, the act of consumption, or the expectation of that act. The ambiguity applies to most terms in economics, but nowhere is it clearer than in the meaning of the word "real" in the term real income. Real sounds as if it refers to the first of the above meanings, emphasizing something tangible, or at least observable. But the intent of the word as a modifier of "income" is to indicate that with this concept we are getting down to measuring the amount of the good in a way that is closely associated with satisfaction generation.

An ambiguity of this kind, embedded deeply into the center of the discipline, would seem to be profoundly unscientific. And so it is. Liberal economists are often entirely unaware of the ambiguity; however, their work tends to push them farther and farther toward assuming the objectivist meaning of the term in designing research and interpreting problems. This produces the trappings of real science, avoids difficult issues, and, as we will argue, generates a highly alienated economic "science" and an economist strongly alienated from his object of study.

Alienation has to do with separation, with a division that sets the individual apart from essential portions of her own life. This in turn stunts the individual by depriving him of the possibility of realizing important potentialities. Our economic example suggests one very broad realm in which this separation occurs in the modern world. An individual consumer finds her purchasing power, his objectified income, increasing steadily. Spending this money on an increasing amount and variety of goods, our consumer finds her sense of satisfaction decreasing, is plagued by a feeling of meaninglessness and anxiety, takes to pills or some other soporific, and in time joins the ranks of the seriously ill—but his real income was growing the whole time. Objective real income has become a totally false measure of subjective real income. As we all know, this experience has been the lot of millions of Americans, and their number seems to be growing rapidly.

The conventional economist must ignore this problem, for it would take a complete restructuring of the "science" to deal with it adequately—and besides, in a certain sense conventional economics has been constructed with the purpose of concealing such facts of life. But on this dimension Marxian economics has a very different story to tell. Very possibly Marx's first substantial intellectual achievement consisted of the adaptation of Hegel's concept of alienation so as to make it a powerful characterization of the social aspects of this separation of humans from their own natures and environments.[1] Unfortunately, her descriptive language is very difficult and he did not discuss the concept in an integrated way in her later work. Nevertheless, the basic concept is built right into the heart of Marxian economics and so has played a central role in the radical tradition. The problem of alienation emerges quite naturally in any radical discussion of contemporary economic problems.[2]

However, having once noted this fact, one must also notice an important difficulty. Alienation is a ubiquitous fact of life under capitalism. Its existence and importance is not at all controversial. But it has not proven to be an easy matter to get at the causes of alienation. Empirical research on the subject is quite spotty and superficial, and indeed there are hardly existent well-developed theories to provide a base from which such empirical research could begin. As a result, there is a considerable range of beliefs about the relationship between the extent of alienation, the nature of society, and the appropriate social processes for disalienation. We illustrate with two relatively extreme views, which may be contrasted with the more or less intermediate position argued in various places in Part I.

Marcuse represents one such extreme.[3] She argues that the social adaptation to industrial society has produced a monstrous and all-embracing depersonalization of human life, and a separation of the individual from almost every central aspect of his being and environment. Marcuse thinks of technical progress as "the instrument of domination." It works in direct ways, separating the individual from the product of her labor and from the real meaning of productive labor, in the classical Marxian sense. But it also separates the individual from his own consumption in very similar ways, depriving acts of consumption of their meaning and converting them into acts of symbolic repression of others or into means to some further end. It permeates political activity. Technological domination deprives the individual of her means of critical appraisal of serious political alternatives, turning the welfare state into "a system of subdued pluralism." It even comes to dominate language, fostering the separating, means/end dichotomy, developing an "overwhelming concreteness" to language that once again deforms and withers the human critical faculty, thus finally producing industrial society's ultimate product, the one-dimensional person.

One of Marcuse's most striking images is that of the "happy consciousness." The one-dimensional human comes to identify with his alienated state; this tends to produce a brittle and superficial happiness that seems never to penetrate beneath the surface of life. The happy consciousness believes that "the real is rational and the system delivers the goods." With this image does Marcuse have, among others, the liberal economist in mind?

Marcuse's argument is presented powerfully, the very striking imagery being most effective in conveying the sense of humans trapped in a clean, bright, chrome and plastic hell. Furthermore, she describes in abstract form instances I am sure each reader can find among his own experiences. The reader of Marcuse will find it very difficult to deny that alienation is a central fact of contemporary existence.

However, many radicals find her account of causation less persuasive. The culprit for Marcuse is the industrial society, which seems to mean any society in which modern industry dominates the forces of production. The overwhelming nature of the problem of alienation, as Marcuse sees it, seems to make attempts at change doomed from the start. At times he seems to say that anyone deeply exposed to industrial society has become too tainted to be effective in combating its consequences. At other times, she seems to feel that through education over a long period of time, enough members of industrial society can escape its thrall to bring about the needed changes. But just what those changes are is quite unclear.

Without denying any of the basic Marcusan images of alienation, I think there are more grounds for hope, perhaps even optimism, than he allows. In the first place, there is the argument that technology is endogenous to society. Thus the alienating technology that confronts us today is in part a product of the relations of production of the past, as argued in chapter 3. The implication of this thesis, that we have considerable ability self-consciously to control the directions of future movement of technological change, is that by changing the relations of production we *can* affect alienation. Second, there is no good reason to place all the blame on the forces of production. Even without a lot of fancy machinery, capitalist society was able effectively to depersonalize human relations; after all, Marx's powerful statement of the problem of alienation was written when the technological features of contemporary industrial society were almost entirely unknown. And finally, not all radicals would want to place as much emphasis on alienation as Marcuse has. They would note that Marcuse was himself not an economist, and that the kind of emphasis she places on this one issue tends to mute the role of the material conditions of production, forces and relations both, as the base from which such superstructure questions should be analyzed.

This last point brings us to our other extreme position, represented by the Yugoslavs. Yugoslav economists tend to have very little to say about alienation. When they do mention the concept, it is usually the objective form they are discussing: By alienated labor they simply mean labor power that has been stripped of control over the means of production with which it is to be combined. But that is not to say that alienation is not taken seriously. Rather, the Yugoslavs seem to have a different view as to how disalienation occurs.[4]

For this school of thought, alienation and power are very closely linked. Product and productive activity alienation result from lack of control of one's environment, and by implication consumption, political, and other more diffuse forms of alienation are similarly a product of lack of power. Hence the problem of disalienation is not one of reeducation but of providing that power to the workers. Workers' management is, of course, the major device proposed by the Yugoslavs in the service of this goal. By returning to the worker genuine influence over the conditions of work and over the nature of the products of her work, he will tend to lose that sense of separation from her environment, to become once again whole. In realizing this aim the worker must learn to understand his factory's alternatives, and learn how effectively to defend her choices within the community of workers. In this way that critical faculty that is the hallmark of Marcuse's two-dimensional human is restored.

Disalienation in this theory is an emergent property of increasing power. But it is important to note that the power must be genuine; that is, it must entail the ability of the workers to really decide for themselves about those major elements of their work environments. And this in turn means that the workers must be properly informed. There must be full access to any information *they* deem necessary in appraising their alternatives. Consequently, disalienation can only develop within a relatively open environment.

The emphasis on disalienation as an emergent phenomenon is important. The suggestion is that there is no privileged group in society that already knows what it is to be disalienated, and that can serve as a vanguard to lead the rest of society in the right direction. Rather, it is argued that the workers will define for themselves their own future states of consciousness, and that those states will be a joint product of the democratic conditions of labor and their own emergent natures. The best society from the point of view of disalienation is the most open and democratic one.

This is an argument with great appeal, and much of it has found its way into the optimal radical economic world view. However, it does seem to ignore some essential problems and uncertainties. One major problem that the above argument—though not in fact the Yugoslavs— ignore is the problem of transition, especially in a world that still contains hostile and powerful capitalist regimes. A second issue it ignores is that of the remnants of previous social structures that survive for a long time after the revolution. In Yugoslavia nationality differences have been perhaps the most persistent of these remnants. And finally, it seems a bit one-sided. In effect it ignores the difficulties raised by Marcuse. The workers by themselves do not have the capacity to design a new and disalienated technology; to do this will require a joint and self-conscious effort by broad segments of many societies. To the extent that the democratic associations of workers are constrained in their choices to essentially alienating technologies, technologies that strongly condition the environment of work, the power to choose has not really been acquired by the worker communities.

Both Marcuse and the—somewhat vulgarized in our account—Yugoslavs have important things to say about alienation. When toned down a bit, they seem more to complement than to contradict one another. But one should remember that we are still very ignorant of the causes and cures of alienation. There are a variety of modes of production operating in the world today. Careful study of alienation as it has historically affected these various societies will hopefully provide us with stronger guidelines for the future. Nevertheless, I doubt that either view described in this chapter will prove to have been fundamentally wrong.

CHAPTER 17

Horvat

BRANKO HORVAT is the most interesting and probably the best known of Yugoslav economists. Though his views are highly original, they are also strongly supportive of the basic thrust of the Yugoslav way to socialism. Horvat is well trained in both Marxism and neoclassical economics. He probably represents his generation's best shot at making tenable an eclectic socialist economics, one that mixes the strong features of both those strands of economic thought to provide an integrated socialist analysis of socialism—and of capitalism too. In this chapter we first outline some of his major arguments and then offer a few comments.[1]

Horvatism

Despite Marx's tremendous intellectual accomplishments, including unusual prescience as to future trends in the development of capitalism, there are two major trends in the twentieth century that were incorrectly or incompletely foreseen by him. The first of these is state capitalism, which may be taken either as the highest stage of capitalism or the lowest stage of socialism. Marx correctly foresaw the rise of large-scale organizations. Indeed, this rise is inevitable, meaning by inevitability that large-scale organizations are both feasible forms of organization and of superior efficiency to the petty capitalist organizations of a century ago. What Marx failed to see was the class conflict that is inherent in bureaucratic systems of control. Once the bureaucracy is established as a dominant system of economic control, a class is created with an interest in preserving its power and relative affluence against outsiders.

As noted above, it makes little difference whether such a society calls itself socialist or capitalist. It does, however, make some difference whether it is essentially one big bureaucracy or not, and whether it developed out of capitalist preconditions. Max Weber pointed correctly both to the efficiency properties of rational bureaucracy and to the importance of rational capitalist calculation as an element in promoting that efficiency. As a consequence of the absence of a developed capitalist rationalism, a bureaucracy imposed on an underdeveloped country will be especially inefficient. However, Ludwig von Mises pointed to a vital aspect of bureaucracy that Weber ignored, namely, the absence of a basis for rational calculation within a large-scale organization that does not contain internal markets. In addition, there are the usual problems of bureaucracy: information absorption and distortion by supercautious middle-level bureaucrats, the aggregative and therefore rough-and-ready nature of the key highly centralized decisions of the top leadership, the tendency to "pass responsibility up and work down," and so on; all of which make very large bureaucracies extremely inefficient. The tendency toward capitalist concentration thus produces at first a higher and more efficient stage of society, but it also contains the seeds of its own destruction as still further growth tends to produce unwieldiness rather than rationality.

The second modification of the classical Marxist picture of social development leads to the assertion of the inevitability of worker control, inevitable again in the sense that it represents a feasible social form whose efficiency is superior to that of the historically preceding form. As to the first aspect of inevitability, there is a clear trend toward increasing worker participation in management. Partly as a consequence of the workers' natural inclinations, there has been some attempt to establish factory committees during every major revolution from 1848 through the Russian and other revolutions around World War I to those of Poland and Hungary of 1956. Partly as a consequence of the needs of state capitalist government, some measure of worker control has been introduced in a number of nationalized industries, and in Great Britain, Germany, and the United States, among others, during wartime. Partly as a consequence of the initiative of enlightened captalists, from profit sharing to production conferences, it has served to increase efficiency and therefore profits. This broad trend strongly suggests both feasibility and superior efficiency. However, it was not likely that full-scale worker control of industry would come first to a developed state capitalist society. What was needed was, first, a revolution to sweep away traditional attitudes and authority in a less-developed country, and then a developing trend toward workers' management that preceded the entrenchment of the new bureaucracy. Hence, Yugoslavia provided the first instance of this higher form of society, which realizes for the first time the ultimate Marxian aim of free associations of workers.

On the morrow of the revolution, the new society faces a number of key policy choices that, implemented over the years, will have a vital impact on the question of whether the political revolution will turn out to have been successful. Of course, in a Third World country one of these choices must be with respect to the tempo of economic development. However, many observers have failed to note that this issue is not one of consumption now vs. consumption later. A well-designed policy can actually do both. The point is that, by devoting rapidly increasing resources to investment, the general productivity of the economy can be increased enough to accommodate both an increase in consumption and the increased allocation of resources to invest-ment. This principle was discovered in the Soviet Union in the twenties and has been widely—if not always efficiently—applied in socialist countries ever since.

Since history operates strongly through efficiency in generating new social systems, the choice of regimes of resource allocation is another crucial decision the new socialist leadership must face. The evidence favors markets as a major resource allocator under socialism. In fact, market processes generate salable levels of output and prices that reflect true levels of relative scarcity better than do bureaucracies, particularly those Third World bureaucracies that have not yet been through their rationalist purging. But, of course, markets cannot be the only planning mechanism. The society's leadership must monitor out-comes and be prepared to intervene with self-consciously planned solu-tions to problems in areas where the market does not, or obviously can-not, produce efficient results.

Consequently, in an optimal socialist regime enterprises will be controlled by the workers, who strive to maximize the profits of the enterprise, receiving a share in the profits as well as wages for their efforts. The profits share represents a return to the workers for their newfound function as entrepreneurs. Since the state, through the plan-ning board, will also be involved in many investment decisions, as well as in setting overall lines of development for the economy, it too is entitled to a share as joint entrepreneur. And since this entrepreneur-ship is diffused throughout a socialist society, no particular workers' collective can claim title to the full profits generated by the market operations of the enterprise.

The ultimate optimal distribution of income will be egalitarian. This, of course, follows from well-known principles of socialist equity; it can also be derived from the basic assumptions of neoclassical economics. However, initially the distribution of income will be in-egalitarian, reflecting the fact that in the earlier stages of a socialist regime the workers continue to have a strong personal material incen-tive. By paying more for more and better work, productivity is en-hanced, and consequently so are both the growth rate and the current rate of consumption of the entire society. However, historical trends show that as output per capita increases, the effects of differential incomes on productivity tend to decrease. So the rule for distribution —maximum equality consistent with maximum production—will, over time, generate a more and more egalitarian current income distribution. At some point the participatory spirit becomes dominant and society begins truly to move beyond the stage of socialism.

There must be a central authority in society, both for defense against external threat and to provide domestic order. In addition, there is a third vital function for such an authority, namely, to provide basic guidelines for the further social development and to provide the educational base for that development. The Communist party is the logical agency to serve as this central authority. However, once the new socialist society has established its basic institutions and sees them functioning effectively, the party must not continue to function as if it were still engaged in a desperate and clandestine struggle to overthrow the government. Since there are many choices to be made and much uncertainty as to which course is best, the party must be an organization within which a continual dialogue takes place. Furthermore, there must be internally democratic decision processes at work, even though hierarchic structure cannot be avoided. Each level of the party must be capable of compelling the next higher level to consider seriously its proposals, and decisions at each level must involve the principle of majority rule. In this way both at the top and the bottom the social institutions are providing models of democratic and participatory decision making to the rest of society.

Commentary

Clearly, there are some very important differences of viewpoint between Horvat on the one hand and Baran and Sweezy on the other. Probably it is reasonable to say that Horvat comes much closer to a view that a conventional economist might find acceptable than do the other authors; perhaps Marcuse represents in some sense the opposite extreme from Horvat. Nevertheless, some portion of the apparent differences is attributable to different emphases rather than straight disagreements. And this in turn can mean that to some extent these authors are complementing one another's work rather than being mutually contradictory.

Consider the case of the treatment of bureaucracy. In Part I we put main emphasis on the mixing of politics and bureaucracy as the generator of waste under monopoly capitalism. Baran and Sweezy emphasize the difficulty in finding a use for the rapidly rising surplus under that regime, despite the existence of very obvious unmet needs. Horvat, in contrast, attacks the inability of a bureaucracy effectively to appraise its alternatives, to measure the net benefits of alternative courses of action, as the principal culprit in generating inefficiency. One notes that these are not mutually contradictory positions. It is quite conceivable, even plausible, that all three are at work simultaneously in producing the appalling mess that we call monopoly capitalism. It would be useful to have a more definitive view as to the relative importance of the three factors, which is of course a task for serious empirical investigation. In this area radical economics is well stocked with plausible hypotheses, and it would seem that the radical position's plausibility would not be significantly affected whichever one turned out to be the more nearly correct one.

The issue of worker control is somewhat different.[2] *All* socialists believe in some form of worker control over their own worklives and over society's productive activities. But there are substantial disagreements as to just how control is to be exercised and over which variables are to be controlled locally and which centrally. The Chinese, for example, subscribe to the vanguard-party thesis, which means that worker control is somewhat indirect. Nevertheless, the Chinese work hard to make the relationship as direct as possible by such devices as having workers and managers perform one another's work for certain periods of time, and by forming joint teams of workers, technicians, and cadres to solve enterprise problems. Horvat wants to put direct control over the key enterprise decision variables—output, working conditions, investment—into the hands of the factory's own workers. However, such worker bodies are still constrained by state policy and framework legislation, and also by the operation of the market, which determines many aspects of the consequences of alternative courses of action.

A difference of the above kind is rather fundamental. Our optimal radical takes the tack that there are many ways to socialism, that one of these may be best in one environment, the other in another; we also argue that just what should be done in the area of worker control is not yet known with full confidence, so that further social experiment is not only in order but should be respected by other socialists. That is especially true since all socialists agree that a major feature of capitalism, namely, the distortions of class, must be substantially eliminated as a precondition for meaningful worker control. Both the above forms of worker control may meet this test.

At a somewhat lower level of generality, there are likely to be disagreements with Horvat's assertion that a regime involving workers' management, in a market environment where each collective tries to maximize its own income, can be very efficient. There is a neoclassical theory that suggests this may not be the case. For example, the fact that in such an enterprise the income shares going to capital and labor are not separated by the operation of markets could cause deviations from efficient use of resources. Just how serious these deviations might be is unknown at the moment. Here once again we have an area of controversy that further experience and study can go a long way to resolve.[3]

But the most serious objection to Horvat's theory is likely to emphasize a different aspect of market socialist behavior. Some socialists argue for the slogan, Beware the market, it is capitalism's Trojan horse! The anarchic market can be very disruptive. But, perhaps more important, the market serves to inculcate selfish and materialistic values, pitting individual collectives against society in a struggle for the surplus. These issues were discussed in chapter 10, but our conclusions were eclectic and tentative. All one can say is that today the disagreement has become somewhat muted, with former "enemies" of the market such as Sweezy and Bettelheim now agreed that it does have some possibly substantial role to play under socialism.[4] That still leaves a substantial gap with respect to Horvat's views.

And so it goes with other issues, such as the role of the planning board, the role of income distribution, and the changing role of the party. Horvat has offered an integrated and thoroughgoing theory of social movement in contemporary state capitalist and socialist society. His ideas have not yet been subjected to a systematic appraisal within the body of radical literature. In this chapter our primary aim has been to show that such an appraisal is very much needed.

CHAPTER 18

Technical Economics vs. Radical Economics?

AS NOTED in chapter 1, radical economics has had a very mixed record of development in the twentieth century. Before World War I it appeared as a rapidly developing new social science. There was a strong underlying core of common beliefs, based primarily on Marx, combined with a very lively development of controversy, the central issues relating to the nature of imperialism, the rise of finance capitalism, and the possibility of an evolutionary development of socialism. But this great promise was not realized in the interwar period. The crisis of monopoly capitalism, the destruction of human life in the war and its aftermath, the rise of a socialist state that was challenged at every turn by its capitalist rivals: These were perhaps the central social factors inhibiting the further development of radical economic science and the radical economic world view. Another factor, as noted earlier, was also of undoubted importance: namely, that monopoly capitalism was lurching from one catastrophe to another while socialism was just getting off the ground in an extremely unfriendly environment; consequently, the fundamental tendencies in both these social forms were not easy to perceive.

It was not until the nineteen fifties that a substantial turnaround began to occur. Key figures in this development in the Anglo-American world were Maurice Dobb and Paul Sweezy, together with Paul Baran. They served both to keep the idea of radical scholarship alive among English-speaking economists and to develop the science further in their own works. But during much of their careers they were rather lonely voices. When once again interest in alternatives to capitalism began to be a serious matter of concern, they provided the base on which further discussion could occur.

The first stage in this new development consisted in the generation of new syntheses, new attempts to characterize the general structure and tendencies in contemporary political economies. The leading names have already been mentioned more than once in these pages, and include Baran, Dobb, Horvat, Mandel, Marcuse and Sweezy. Their works of synthesis essentially were a set of competing economic world views, posing for radical economics the question of the extent to which they could be reconciled once again, as had been the case a half century earlier, to a common core of belief. It must be admitted, however, that these works, despite their obvious differences, did not generate much in the way of attempts at appraisal and synthesis. To my knowledge, Part I of this work is the first attempt to do this, though hopefully not the last.

The next stage, naturally enough, consists in developing a radical scholarship that attempts to explore in detail various relevant aspects of the structure and tendencies of political economies. Efforts at appraising the rather general factual assertions that appear in economic world views are a part of this. So is effort devoted to resolving differences among competing radical world views. So is the development of a serious running critique of liberal economics, both at the methodological and the substantive levels. Clearly, the carrying out of each of these types of activity requires the joint efforts of a considerable number of radical economic scholars. What was still not feasible even in the sixties has become possible today, for trained radical economists now exist, perhaps for the first time, in sufficient numbers and under conditions that are not inconsistent with the carrying out of this program of scholarly work.

However, there is still an argument that questions the desirability of taking this step. It has several parts, and we react differently to these parts, so we will take them one at a time. First there is the argument that the revolution cannot wait for such nit-picking activities; making it happen requires the full-time efforts of all radicals. This seems to have been a widespread feeling among younger radicals during the sixties, but has waned considerably. The change in part reflects a recognition that the revolution may not come to many monopoly capitalist countries for a long time, but perhaps it has come mostly from a growing recognition that what happens after the revolution is connected in important ways to what happens before. Revolutions can go astray, and especially the leadership can lose contact with the masses. If difficulties such as those that have beset the Soviet Union are not to happen elsewhere, two things must be done. First, the masses must be prepared for revolution; they must understand its purpose and as many as possible of them must participate in bringing it about. Second, the leadership must understand that it is the representative of the masses, which means that it is not beyond the reach of democratic processes. Bringing about such a revolution is a complex matter and requires, among other things, that radical scholarship must be able to make its case effectively, in an environment of serious intellectual struggle with its liberal rival. On this ground the future of radical scholarship now seems secure.

A second argument suggests that radical scholarship, having a function very much different from conventional scholarship, will also have a different future history. The basic function of radical scholarship, so this argument goes, is to understand the structure and tendencies of monopoly capitalism with a view to eradicating it. But this implies that radical scholarship will lose its meaning on the morrow of the revolution. Its basic function is to put itself out of business. Such a position implies much more emphasis on political tactics rather than detailed scholarship as the primary object of study.

There are two objections to this orientation. The first is that it implies that all revolutions will be successful, or at least that the successful ones can be recognized when they occur; that sounds a bit utopian. The second, and related, objection is that it assigns a rather small significance to an intellectual movement that has had a very profound and worldwide impact over the past century. It seems more likely that radical economics has a general perspective toward the investigation of fundamental properties of political economies that, at least in part, transcends medium-term transformations of the relations of production. Indeed, I would be inclined to argue that one of the main potential supports to continuing the initial success of a revolution lies in the existence of a body of radical political economists who can continue to appraise the structure and tendencies of the system.

A third argument raises more complex issues and will occupy the remainder of this chapter. It holds that the problem with radical scholarships lies not in its purely temporary significance but in its class bias. To the extent that it embraces neoclassical economics, it is tainted with bourgeois values and elitism. To the extent that it relies on technical Marxism it, together with most liberal economics, is divorced from the world of reality. And to the extent that it apes the nit-picking of contemporary historical scholarship, it preserves "reality" at the cost of triviality. There is, I believe, some truth to each of these charges. Let us take them up in order.

The first problem with neoclassical economics is its elitism. This is inherent in the method. It presumes that there is a specialist group, the master of its arcane language and techniques, who are the only ones who can successfully solve economic problems. It is especially well adapted to serving the needs of bureaucrats, and has been spreading like wildfire in recent years as a technique applied throughout bureaucracies. Socialist societies have found their own bureaucratic classes split on the question of using neoclassical economics, but the more "progressive," i.e., technocratic among them tend unequivocally to support these techniques. It gives the bureaucracy a tremendous advantage in dealing with outsiders, such as the masses, simply because of its highly technical nature, and the large costs that must be incurred to generate a "scientific" result.[1]

If this were the only way to get at the truth about how economies work and how to design policies, then we would all be in the freedom-is-the-appreciation-of-necessity box together. But that is just not the case. In fact, much of the technical side of economics is pure mystification, self-serving to the affluent and influential economics profession. Relatively simple techniques, which can be understood by very broad segments of the population, suffice to support nearly all of our current genuine knowledge about how economies work. Further efforts could expand this demystified realm even further. For example, the Chinese at one point were teaching "people's operations research" to the peasantry. The idea was to get peasants to think in elementary quantitative terms about the cost and productivity of alternative methods of plowing or planting or machine use. A certain amount of effort spent in obtaining relevant numbers, plus a few simple calculatons, sufficed to give peasants who typically had no more than a primary education the full basis for making their own decisions in these areas of central collective concern. By recognizing that in economics "crude is beautiful, sophisticated is bourgeois," an important element of participation, and no doubt of disalienation, was introduced into peasant life. Hopefully that movement will spread.

A second problem with neoclassical economics is that its technical structure fundamentally reflects its philosophical origins. It is the science of society of the rising bourgeoisie. As such it assumes right at its heart that individuals are what count and that the relations of production are thoroughly privatized. These assumptions come in through price theory. The efficiency properties of neoclassical "solutions" to economic problems are inevitably based on price theory. For example, a fundamental assumption of the theory of consumer behavior is that one person, or family, or consumption unit's satisfaction from a particular consumption package is independent of the satisfaction of other consumption units. Exit socialism right there! The corresponding assumption with respect to production, which has production carried out in atomized units, flies in the face of the trend, already noted by Marx and consequently more than a century old, toward increasing socialization of the forces of production. It is this assumption that leads to the neoclassical economist's helplessness in the face of ecological and environmental problems. Neoclassical economics is both unsocialist and unhistorical in more than one sense of each word.

Neoclassical economics has *some* helpful things to say about how societies work. Mathematics and statistics, as is now almost universally recognized by radical economists, are not apologetic wherever and whenever they are used. The problem is one of emphasis and of the nature of the particular use to which these tools are put. But controlling the tools has proven to be an extremely difficult job. Unfortunately, even though they are irrelevant, mathematical games are, like chess, fun to play. In American academia playing these fun games is, in addition, the best way to have a successful career. These pressures are very insidious and are felt by radical as well as by conventional economists who are caught up in the atmosphere of contemporary academe. Recently, for example, there has been a substantial proliferation of mathematical Marxism. Much of this literature is designed to show that models derived from Marx's theories of value and of expanded reproduction can be brought to bear on contemporary economic puzzles as effectively as corresponding neoclassical models. Some of this work has served the useful purpose of exposing difficulties in the neoclassical models. Some of it has usefully criticized the narrowness of those models, as in debunking the notion that capital is a separate and homogeneous factor of production, or that production and distribution can be separated in analyzing economic growth. However, these models are in fact profoundly un-Marxist. They tend to accept the questions asked by bourgeois economists as the starting point for their development. This usually condemns them to irrelevance, as well as to elitist mystifying of the historical processes that generate economic development. Development of this literature distracts radical economists from their central task of developing a pragmatic and intelligible economics that will grasp the dynamic and historical nature of the contemporary crisis and the forces making for further change.

There is a third kind of trap that radical scholars can fall into. This one, perhaps surprisingly, is Marxist exegetics. Marx continues to be an author whom every serious radical economist will have read carefully. Understanding Marx for a Marxist also means understanding the historical context in which he developed his ideas. In the same way, a radical economist will feel obliged to acquire an understanding of the development of political economic thought in the twentieth century as an aid in grasping the forces at work on contemporary intellectuals. But here the radical economist must once again be careful to find the right line. Too much of too many radical journals is devoted to painstaking analysis of aspects of Marx's thought, or of that of leading Marxists of the past, without any effort being made to relate the arguments to the real problems of the twentieth century. This too it seems can be a pleasant and escapist game to play, distracting the analyst once again from the main line of effort.

A second difficulty can arise if one begins to worry about whether a particular line of argument is really Marxist or not. In many socialist countries economists have not been able effectively to deal with problems of economic valuation because they are constrained to a pricing schema that can be found somewhere in *Capital*. This misconceives Marx's own notions as to the course of development of ideas and so is self-contradictory. It also suffers from the fact, argued in chapters 5 and 6 above, that Marx's value theory is no longer an adequate characterization of the value problem in either monopoly capitalism or socialism.

These difficulties have hampered the development of radical economics in the twentieth century and have no doubt contributed to feelings that radical scholarship was perhaps unnecessary. But all they really demonstrate is that radical scholarship is difficult. Operating in a relatively hostile environment in which the structure of incentives and the intellectual milieu combine to encourage elitism and game playing, the radical scholar must have a very firm sense of purpose if she is to succeed in making a useful intellectual contribution. A well-developed radical economic world view that provides a coherent, inclusive, and integrated appraisal of the central economic issues of our time could be helpful in affirming that sense of purpose.[2]

CHAPTER 19

The Role of Revolution

ON THE LEFT, attitudes toward revolution have probably been converging in recent years. It has become increasingly apparent that Third World countries will not be allowed to embrace socialism by an evolutionary process mediated by parliamentary democracy. Violent revolution, as the Chinese are always saying, seems unavoidable as an instrument of the transition. This is not a matter of preference by the left; rather, it is an alternative chosen for the left by the forces representing monopoly capital. If the latter could establish their willingness to let such things as referenda and genuinely representative parliaments decide periodically whether to go the socialist way or not, a very large segment of the left would immediately drop its revolutionary orientation. But as was argued in Part I, this alternative is, if anything, less likely to occur now than before. Revolution is the only alternative to continued massive misery and oppression.

However, the social democratic deviation has not disappeared and at times can be a very powerful lure to leftists in one country or another. The key ingredient of this deviation is its promise of jobs and influence to the leadership via parliamentary participation. Social democratic parties retain their formal identities, and they often continue for years to use the revolutionary language of their earlier days. But parliamentary participation is a true Trojan horse for a revolutionary party. In examples from Germany to Chile, it has shown its ability to erode revolutionary interest and organization, its ability to inculcate the art of parliamentary deal making and its corollary, not rocking the boat. When the crunch comes the starch is gone; the crunch came in Germany in the early thirties and in Chile in the early seventies, and the outcomes were very similar. But the seeds of failure were in both cases sown over a period of decades of participation by the leadership in the local parliament, where the only real road open to them was that of loyal opposition.

However, the problem of revolution has become much more complicated in the homelands of monopoly capital. In all of the most advanced countries, there is a strong "democratic" left and a weak revolutionary left. There is relatively little support among the mass of the urban workers for a revolutionary program, and the peasantry tends to be relatively small and affluent. In the United States the existence of a powerful military, trained since the sixties in the arts of "riot control," complicates the problem still further, as does the existence and possible irrational use of nuclear weapons. As noted in chapter 11, there is no obvious way to describe feasible paths to revolution in these countries.

It was once fashionable to characterize stages in the development of a revolution. By such an accounting the United States is presently in a rather quiescent state. At such times the emphasis tends to be placed on movement building rather than on direct action. Some writers have suggested that it may take decades of movement building for success to become a visible prospect. Others have emphasized the state of objective crisis in which monopoly capital finds itself, a state that shows every promise of continuing and at times intensifying. For the latter emphasis, clearly there is an important immediate goal: building a movement capable of acting should a crisis opportunity emerge in the near future.

Attractive as such an approach must seem, we have argued against it, primarily on the ground that the left is not prepared intellectually for the very complex demands that revolution and its aftermath pose in monopoly capitalist countries. Programs of the splinter groups currently active run from support for institutions similar to those in each of the major existing socialist countries to a variety of policy platforms whose implementation is both dubious and essentially undefended. The conditions of revolution and the splintered state of the left would not permit any reasoned or democratic resolution of these differences. At least the groundwork for that process of resolution has to be laid first. And then there is the consideration, given great emphasis in many segments of the American left, that democracy must be an integral part of socialism in America. Surely this means that a substantial segment of the population must come to recognize the advantages of dramatic structural change of the American economy well before the revolution.

Can such an educational effort be carried out without producing essentially the social democratic deviation? Won't institutions such as schools and universities and other potential workplaces for radicals have much the same effect on their values as would participation in a legislature? Certainly it is a risk, but what is the alternative? The success of revolution in America depends on the failure of monopoly capitalism in America. If the latter doesn't happen, then the former surely won't. If revolution is going to happen, it will cast its shadow well beforehand. A successful movement will have to be one that can perceive the shadow, interpret it persuasively, and offer some realistic vision of a better world. Opinions can certainly differ as to how best to carry out such an effort. But if opinions continue to differ so substantially and bitterly over the program, the policy, then the effort is probably doomed to failure.

A central message of this book is that from the economic side there appears to be a firm base for the development of substantial unity of intellectual orientation on the revolutionary left. This base is rooted in Marxism, but involves recognition that much of Marx's analysis needs substantial revision before it can interpret successfully the events of the twentieth century. It is based on a belief that successful revolutions can occur, that they have occurred, and have transformed whole societies. It is based on a belief that there are still, as far as we can tell, many paths to socialism, and so is based on tolerance of diverse revolutionary experiments around the world. And it is based on a belief in the need for political unity in the revolutionary left within each of the advanced countries of monopoly capitalism as a precondition for success. But for that to happen there must be, to put it bluntly, further study.

Notes*

Chapter 1

1. Paul Baran, *The Political Economy of Growth* (New York: Monthly Review, 1957).* The second edition, dating from 1962, contains reactions to reviews of the book and further comments, especially on socialism.
2. Paul Baran and Paul Sweezy, *Monopoly Capital* (New York: Monthly Review, 1966).*
3. Ernest Mandel, *Marxist Economic Theory*, 2 vols. (New York: Monthly Review, 1968).* The French original was published in 1962.
4. Ernest Mandel, *Late Capitalism* (New York: Monthly Review, 1976).
5. Branko Horvat, *Toward a Theory of Planned Economy* (Belgrade: Yugoslav Institute for Economic Research, 1968). The Serbo-Croatian original appeared in 1964, and reportedly was completed several years earlier.
6. Our list, of course, does not contain the names of all seminal contemporary radical writers. In particular, the names of Jürgen Habermas, Jean Althusser, and Charles Bettelheim are missing, the first two because their contributions are primarily methodological in nature, the third because he underwent a dramatic change in orientation in the sixties. A discussion of methodological issues is beyond the terms of reference of this work, which concentrates on arguments that contribute more directly to the understanding of contemporary society. Bettelheim's name will crop up in later chapters.
7. George Wang, ed., *Fundamentals of Political Economy* (White Plains, N.Y.: Sharpe, 1977).
8. Mao Tse-tung, *Selected Works* (Peking: Foreign Languages Press, 1961–1977).
9. John Gerassi, *Venceremos* (New York: Clarion, 1968).*
10. Clearly, the time has come to eliminate an anachronism from the English language, namely, the mandatory sexual distinction when using personal pronouns. In an effort to hasten the process of elimination, I shall alternate the use of the feminine she/her with the masculine he/him throughout this book.

A brief argument in defense of this procedure is in order, even though this is by no means the first time it has been used. Given the structure of American society a century or a century and a half ago, it would have been quite convenient, certainly for whites and possibly even for blacks, to have pronouns that distinguished individuals and groups by race. The society was thoroughly permeated by racism and so it would have been "natural" for this to have become embedded in the structure of language. This did not happen because the morphology of language tends to change slowly, much more slowly than have societies in the era of the industrial revolution. But had such pronouns emerged then, radicals would unquestionably have long ago found their continued use intolerable. The subservience of women began much earlier so that there was plenty of time for the pronominal distinction of sex to become deeply embedded in the language. But surely the reasons for eliminating that distinction are as pressing as are those for the elimination of racism.

There remains the question of the specific manner of dealing with the problem adopted in this book. The first thing to note is that these days it is rarely an important distinction semantically; for example, in no single instance in the present work do ambiguities of meaning result from the assumption that all pronouns used are unspecific with respect to sex. Second, I ask the reader to bear with the usage through a reading of the book before drawing conclusions. More than one reader of the prepublication text found that by the time he had finished the text the usage no longer seemed particularly strange to her. It is a simple and effective device and involves no annoying search for circumlocutions.

* Starred items are available in paperback edition.

Chapter 2

1. Richard Edwards, Michael Reich, and Thomas Weisskopf, eds., *The Capitalist System, A Radical Analysis of American Society,* 2nd ed. (Englewood Cliffs, N.J.: Prentice-Hall, 1972) * is a collection of readings organized along the same lines as the present chapter.

2. A statistical picture of inequality and of the poverty, malnutrition, and even hunger that are a common and persistent fact of life among poorer Americans can be found in a pamphlet by Letitia Upton and Nancy Lyons, *Basic Facts: Distribution of Personal Income and Wealth in the U.S.* (Cambridge, Mass.: Cambridge Institute, 1972) and in *Ten-State Nutrition Survey* (Washington, D.C.: U.S. Government Printing Office, 1972). For a useful descriptive and statistical survey of inequality, see Herman P. Miller, *Rich Man Poor Man* (New York: Crowell, 1971).

3. Joseph Kershaw, *Government Against Poverty* (Washington, D.C.: Brookings Institution, 1970).

4. The figure is a guesstimate by the United Nations' Food and Agriculture Organization. As reported in Alan Berg, *The Nutrition Factor* (Washington, D.C.: Brookings Institution, 1973),* chap. 1, the still very high death rates for infants and small children in Third World countries are mostly a product of malnutrition.

5. Simon Kuznets, *Modern Economic Growth, Rate, Structure and Spread* (New Haven: Yale University Press, 1966),* p. 423.

6. A useful survey of American imperial foreign policy can be found in William A. Williams, *The Tragedy of American Diplomacy* (New York: Delta, 1962); see also Franz Schurmann, *The Logic of World Power* (New York: Random House, 1974).*

7. See U.S. Arms Control and Disarmament Agency, *World Military Spending,* an annual publication in recent years. Actually the figure had already reached $216 billion in 1971.

8. Two useful works on this subject are Raymond Franklin and Solomon Resnick, *The Political Economy of Racism* (New York: Holt, 1973),* and Victor Perlo, *Economics of Racism USA* (New York: International, 1975). The facts of women's subservient status in the American economy are described in U.S. Bureau of the Census, *A Statistical Portrait of Women in the U.S.,* Special Studies Series P-23, no. 58, 1976.

9. Franz Fanon, *The Wretched of the Earth* (New York: Grove, 1968),* p. 32.

10. The concept of subdued conflict is developed in Herbert Marcuse, *One-Dimensional Man* (Boston: Beacon, 1964),* esp. chap. 1–4.

11. There are two ways to demonstrate the profound irrationality of capitalist society. Novels often capture the human truth of a dying and corrupt social order most effectively. Works that inspired my generation include Robert Briffault, *Europa* (New York: Scribner's, 1935), dealing with the European ruling classes before World War I; Martin Anderson Nexö, *Ditte,* 3 vols. (New York: Holt, 1920–22), the story of a working-class woman's struggle against oppression in Northern Europe; Andre Malraux's two novels of war and revolution, *Man's Fate* (New York: Smith and Haas, 1934), dealing with the Chinese revolution, and *Man's Hope* (New York: Random House, 1938), dealing with the Spanish Civil War; John Steinbeck, *In Dubious Battle* (New York: Modern Library, 1936), the story of a great California farm workers' strike in the thirties; and Henrik Ibsen's *A Doll's House* (Boston: Luce, 1909). A second way is to contrast the reality we experience with the way a more rational social order functions. This contrast, of course, cannot yet be made for the United States, but it *is* possible to do so for China, where victims of the old society have recounted their experiences before and after the revolution. Of special interest are Jan Myrdal's *Report from a Chinese Village* (New York: Random House, 1965) * and William Hinton's *Fanshen* (New York: Random House, 1966),* the latter also showing the detailed process of revolutionary social transformation in a North China village.

Chapter 3

1. A fascinating and lucid account of the development of socialist thought in interaction with the development of capitalist and socialist societies is provided by John Gurley, *Challengers to Capitalism* (San Francisco: Chandler, 1976).*

2. Anyone who wishes to understand contemporary radical economic thought must know something about Marx's theories. An extremely lucid and simple introductory survey was produced by John Eaton, *Political Economy, A Marxist Text-*

book (New York: International, 1963).* Perhaps the best English-language account of Marxian economics remains Paul Sweezy's *Theory of Capitalist Development* (New York: Monthly Review, 1942).* These works should not discourage the reader from looking into Marx himself; many of Marx's and Engels's most interesting writings are available in the paperback *Selected Works* volume put out by International Publishers (New York: New World, 1968).* A fascinating and still useful account of Marxism-Leninism is by the two Russian revolutionaries, Nikolai Bukharin and Evgeniy Preobrazhensky, *The ABC of Communism* (Baltimore: Penguin, 1969),* originally published in English in 1922.

3. In the century since Marx a great many historians have been laboring in archives and digs, producing new information about the human past. Some of this has required a revision of Marxian ideas, though his fundamental principles of historical tendencies have not been affected by this mountain of research. The new Marxian interpretation of early history is presented in very readable form in the works of V. Gordon Childe, especially her *What Happened in History?* (London: Pelican, 1954).* The transition from feudalism to capitalism was intensively studied in Maurice Dobb's seminal *Studies in the Development of Capitalism*, rev. ed. (New York: International, 1963),* and very good accounts of key aspects of the industrial revolution have been provided by Eric Hobsbawm's *Industry and Empire* (London: Pelican, 1969) and *The Age of Revolution 1789–1848* (New York: Mentor, 1964). Much of the reinterpretation has been put together in a grand synthesis of Marxian theory and the new Marxist history in the first volume of Mandel's *Marxist Economic Theory* (New York: Monthly Review, 1968). The radical story of American economic development has been recently told in Douglas Dowd's *The Twisted Dream* (Cambridge, Mass.: Winthrop, 1974). A powerful defense of the thesis that relations of production, not technological imperatives, were the primary influences on the rise of the factory system can be found in Stephen Marglin's "What Do Bosses Do?" *Review of Radical Political Economics* 7 (1974–75).

4. A beautiful set of stories of life in the period of capitalism's great end-of-the-century crisis is to be found in Barbara Tuchman's *The Proud Tower* (New York: Macmillan, 1966).* The "climacteric" thesis was first proposed by E. H. Phelps Brown and Handfield-Jones, "The Climacteric of the 1890s: A Study in the Expanding Economy," *Oxford Economic Papers*, 1952, and has been the source of much controversy since. This era stimulated some major controversies among Marxists, of which perhaps the most relevant for the present chapter was that between the "forces of production" and the "relations of production" proponents. The former believed that the Marxian schema of successive stages of transformation of society was quite rigidly forced on societies so that, for example, czarist Russia would have to develop into a fully capitalist society before socialism could become a serious possibility. The latter, of whom Lenin was the best-known figure, argued that a socialist revolution was indeed feasible in an economically backward country. Variants of this important controversy have continued down to the present day, where, for example, in China Liu Shao Ch'i seemed to support a forces-of-production point of view in his policy of primary emphasis on the industrial transformation of China. Mao Tse-tung rejected this notion and in addition has argued that there is an important line of causation in revolutionary China running from the relations of production back to the forces of production, this being a major theoretical argument in his support of the cultural revolution. The consequent interpretation of world history has been outlined in *The History of Social Development* (Shanghai: Shanghai People's Publishing House, 1974), which will soon appear in English.

Chapter 4

1. Revisionist American historians have recently produced a new analysis of the emergence of monopoly capitalism out of the turn-of-the-century crisis. Of particular interest are Gabriel Kolko, *The Triumph of Conservatism, A Reinterpretation of American History, 1900–1916* (Chicago: Quadrangle, 1967),* and James Weinstein, *The Corporate Ideal in the Liberal State, 1900–1918* (Boston: Beacon, 1968). Branko Horvat's *Toward a Theory of Planned Economy*, as well as his *Essay on the Yugoslav Economy* (New York: IASP, 1969), contain a somewhat different theory. Instead of emphasizing the role of individuals in making key structural decisions in the Kolko-Weinstein vein, Horvat sees the emergence as a product of impersonal forces built into the structure and dynamic of early capitalism; he also finds very similar processes at work in the Soviet Union.

2. The list actually totals fifty-four just to the end of the sixties. It is provided

in Lincoln Bloomfield and Amelia Leiss, *Controlling Small Wars, A Strategy for the 70s* (New York: Knopf, 1969), Appendix C.

3. Pierre Jalée, *The Third World in the World Economy* (New York: Monthly Review, 1969), deals with one of the central processes of twentieth-century capitalism, namely, the emergence of the Third World to economic and political prominence and its interaction with developed capitalism. The emergence of the economic structures of European monopoly capitalism in response to crisis is described in several chapters of the *Fontana Economic History of Europe*, edited by Carlo Cipolla, vol. 5, *The Twentieth Century* (London: Collins, 1976). The dimensions of the current crisis are spelled out in the readings in Richard Edwards, Michael Reich, and Thomas Weisskopf, eds., *The Capitalist System, A Radical Analysis of American Society*, 2nd ed. (Englewood Cliffs, N.J.: Prentice-Hall, 1978), and indeed in the pages of almost every issue of the daily press.

4. For historical surveys of the labor issue, see David Gordon, *Theories of Poverty and Underemployment* (Boston: Heath, 1972),* and Harry Braverman, *Labor and Monopoly Capital, The Degradation of Work in the Twentieth Century* (New York: Monthly Review, 1974).

5. According to the official series, which substantially understates the actual magnitude, unemployment in 1938 was again over the 10 million mark, a 35 percent increase over the previous year; in 1939 it was well above the level of three years earlier. *Historical Statistics, of the United States*, U.S. Department of Commerce, 1960, p. 70.

Chapter 5

1. See Paul Mantoux, *The Industrial Revolution in the Eighteenth Century*, rev. ed. (London: Cape, 1961), for a number of harrowing accounts of the consequences of this depersonalization of human relations.

2. The World War II example was described by Baran in *The Political Economy of Growth*, p. 41. Additional data needed to produce the conclusion in the text was provided by the *Historical Statistics of the United States*, *passim*. Crudely, the calculation is based on two considerations: (1) During World War II only half the total labor force was required to produce a decent standard of living for the American population; (2) productivity per worker has doubled since then. So even after some limited refinement the estimate in the text is quite conservative.

3. See Baran and Sweezy, *Monopoly Capital*, pp. 135–38.

4. Seymour Harris, *The Economics of American Medicine* (New York: Macmillan, 1964), p. 7.

5. For a study suggesting that that paragon of competitive capitalism, the American market for wheat, has an extremely costly information system, see P. L. Schmidbauer, "Information and Communications Requirements of the Wheat Market: An Example of a Competitive System," *Technical Report No. 21*, Center for Research in Management Science, Berkeley: University of California Press, 1966).

6. For references see note 2 of chapter 2.

7. According to the 1976 *Economic Report of the President*, Table B-29, in 1965 "average spendable weekly earnings in private nonagricultural" employment, in 1967 dollars, were $91.30; ten years later they were down slightly to $90.50. Since then inflation has eroded money-wage gains to preserve the stagnation in real terms.

8. For a historical account and analysis, see Marilyn Goldberg "*Housework as a Productive Activity*" (Ph.D. diss., University of California at Berkeley, 1977).

9. For a penetrating account of the development of routinization in American work, see Harry Braverman, *Labor and Monopoly Capital, The Degradation of Work in the Twentieth Century* (New York: Monthly Review, 1974). See also below, chapter 14.

10. In 1974 white infants died during their first year at a rate of 14.8 per thousand live births; "Negro and other" at a rate of 24.9. Maternal deaths follow the same pattern but even more strongly: 10.0 for white women, 35.1 for other races. Data from *U.S. Statistical Abstract* for 1976, p. 64.

11. For a demonstration of the scandalous nature of the official treatment of the concept of unemployment, see Bertram Gross and Stanley Moss, "Real Unemployment is Much Higher Than They Say" in David Mermelstein, ed., *The Economic Crisis Reader* (New York: Random House, 1975),* pp. 32–37. Gross is a recognized expert on the subject. He estimates that the recent real rate of unemployment is about one quarter of our labor force.

12. However, it is widely believed by radical economists to require modification to fit the era of monopoly capitalism. Paul Baran's approach, developed in the two books cited in footnote 1 of chapter 1, was to estimate the size of the surplus generated in capitalist society and to show how that surplus is misused, or lies unused. Joseph Phillips made estimates that are reported at the end of *Monopoly Capital;* however, these estimates need revision. For example, they lump all profits together, though clearly it is the profits of the large corporations that represent the keys to power and influence these days, and profits of the large corporations have been increasing far more rapidly than total profits. See note 7 of chapter 15 below.

However, a fully developed and empirically applicable theory of exploitation under monopoly capitalism has yet to be worked out. Until it is, the workings of exploitation must be studied piecemeal. Among important recent writings that contribute to an understanding of exploitation are: James O'Connor, *The Fiscal Crisis of the State* (New York: St. Martins, 1973),* who provides a theory about several exploitative mechanisms for extracting and allocating surplus through the state; David Gordon, *Theories of Poverty and Underemployment* (Boston: Heath, 1972), who provides an analytical sketch of the two labor markets generated under monopoly capital to match the corporate and market sectors; and Herbert Gintis and Samuel Bowles, *Schooling in Capitalist America* (New York: Basic Books, 1976), who pin down the ways in which fundamental class biases structure American education. Ralph Nader's books provide much detailed information on the operation of the system; an interesting demonstration as to how economic power and political power go hand in hand can be found in the Nader group's recent *The Corporation State, The Monopoly Makers,* Mark J. Green, ed., (New York: Grossman, 1973).* That even so small-scale an institution as the small claims court is immediately subverted from its announced purpose by the system is documented in Dennis O. Flynn, "Reno Small Claims Court: Its Purpose and Performance," University of Nevada, Research Report #10, Bureau of Business and Economic Research, 1973.

Analysis of the class structure of monopoly capitalist society has been the subject of much recent radical discussion and research. T. B. Bottomore, *Classes in Modern Society* (New York: Pantheon, 1966), points out that a class and an elite are two different things, and that contemporary capitalism is still under class control. Some properties of America's ruling class are described in G. William Domhoff's *Who Rules America* (Englewood Cliffs, N.J.: Prentice-Hall, 1967) * and *The Higher Circles* (New York: Vintage, 1970).* The thesis that corporate America is essentially controlled by a managerial group has been accepted by some radical theorists, such as Baran and Sweezy in their *Monopoly Capital* (see esp. chap. 2). The principal opposing view, that control has largely passed into the hands of a much smaller group who, operating through a handful of banks and other financial institutions, are able to control basic decisions for the entire economy is ably defended by David Kotz, *The Role of Financial Institutions in the Control of Large Nonfinancial Corporations,* (Berkeley: University of California Press, 1977). This is one of the most interesting of the controversies among radicals these days, and has important consequences for the understanding of the nature of the contemporary crisis. See also chapters 7 and 15 below.

The issue of alienation raises fundamental questions as to the nature of man and of her responses to changes in the social structure; see chapter 16 below, and the note to that chapter, for further discussion.

Chapter 6

1. *Historical Statistics of the United States,* U.S. Department of Commerce, 1960, pp. 719, 720. Three years after the end of the Vietnam war major national security expenses were up 20 percent in dollar terms.

2. Herbert Gintis and Samuel Bowles, *Schooling in Capitalist America* (New York: Basic Books, 1976). See also Alexander Field's 1974 University of California at Berkeley dissertation, *"Educational Reform and Manufacturing Development in Mid-nineteenth Century Massachusetts."*

3. In 1975 the Department of Health Education and Welfare employed 147,100 persons, while state and local governmental employees in public welfare agencies numbered 339,400. See *U.S. Statistical Abstract* for 1976, pp. 249, 286.

4. *U.S. Statistical Abstract,* p. 520.

5. These quotes were obtained by Leonard Silk of the *New York Times* staff and his associate, David Vogel, and reported in their book, *Ethics and Profits, The Crisis of Confidence in American Business* (New York: Simon and Schuster, 1976).

Chapter 7

1. Naturally, all broad-gauge radical writers have dealt with the question of instability under monopoly capitalism. Two major lines of argument emerged in recent decades. The first argued that the Keynesian revolution had made it technically possible to use the fiscal and monetary powers of big government to control economic fluctuations. However, it was also argued that effective control was not politically feasible because powerful segments of the capitalist class were bound to be hurt in the short run by any effective control policy (high interest rates, for example, mean more profits for bankers but disaster for building contractors). Paul Sweezy has been a major representative of this school; in *Monopoly Capital* the crisis tends to emerge in the political context of increasing and generalized dismay by the populace at the increasing irrationality of resource allocation in American capitalism.

A second line of analysis does not dispute the above arguments regarding the political and structural deficiencies of the system, but adds the important claim that there are fundamental instabilities built into the economic structure. The recent failure of economic control policies carried out by governments that were listening fairly carefully to conventional economic policy advisors has lent credence to these arguments. However, such analyses have not yet become fully developed or integrated with other aspects of radical economic thought.

2. This is the conclusion of both J. C. R. Dow, *The Management of the British Economy, 1945–60* (Cambridge: Cambridge University Press, 1964), and C. D. Cohen, *British Economic Policy, 1960–69* (London: Butterworths, 1971).

3. See Raford Boddy and J. Crotty, "Class Conflict and Macro-Policy: The Political Business Cycle," *Review of Radical Political Economics* 7 (1975).

4. Ibid.

5. See Fred Block, *The Origins of International Economic Disorder* (Berkeley: University of California Press, 1977).

6. See Anthony Sampson, *The Arms Bazaar* (New York: Viking, 1977).

7. For a characterization of the failures of conventional market theory, see chap. 2 of my *Socialist Economy* (New York: Random House, 1967).

8. Two recent University of California at Berkeley dissertations explore various aspects of the instability of Keynesian economic models: J. C. Benassy, "Disequilibrium Theory," 1973 and Alan Shelly, "A Study in Disequilibrium Macroeconomics," 1974. Victor Perlo, *The Unstable Economy* (New York: International, 1973), offers a more intuitive and institutional account of instability in the American economy. But perhaps the best evidence for this thesis lies in the extremely destructive criticism of Keynesian economics that has been developed more from the right than the left, and of which Axel Leijonhufvud's *Keynesian Economics and the Economics of Keynes* (New York: Oxford University Press, 1968) was an important contribution. At the moment it seems that conservative and liberal economists have each succeeded in destroying the policy arguments of the other side, but neither has a plausible and constructive proposal to put in the place of the defeated position.

9. In an interesting recent paper, "Stagflation and the Political Economy of Decadent Monopoly Capitalism," *Monthly Review,* 28 (October 1976), 14–29, Douglas Dowd has argued that the sprouts of our current crisis were growing rapidly during the relatively stable nineteen fifties.

A useful account of business fluctuations, with references, is Howard Sherman's *Radical Political Economy, Capitalism and Socialism from a Marxist-Humanist Perspective* (New York: Basic Books, 1972), especially chap. 7–9 and 17. Various aspects of the crisis of the seventies are discussed in the URPE collection, *Radical Perspectives on the Economic Crisis of Monopoly Capitalism* (New York: URPE-PEA, 1975),* and in Joyce Kolko, *America and the Crisis of World Capitalism* (Boston: Beacon, 1974).*

Chapter 8

1. Though no radical himself, Peter Drucker emphasizes this alarming fact in his characterization of the problem of underdeveloped countries; see his *Age of Discontinuity* (New York: Harper and Row, 1969).

2. Paul Baran's *Political Economy of Growth* * remains the primary work that has come to grips with the nature of underdevelopment and the causes of its persistence in the modern world. Andre G. Frank, in his *Capitalism and Underdevelopment in Latin America* (New York: Monthly Review, 1967), and *Latin America:*

Underdevelopment or Revolution, (New York: Monthly Review, 1969),* has told
the more detailed story of the creation of underdevelopment in Latin America. Carl
Riskin, "Surplus and Stagnation in Modern China" in Dwight Perkins, ed., *China's
Modern Economy in Historical Perspective* (Stanford, Calif.: Stanford University
Press, 1975), shows for the Chinese case just how large the relative surplus was—
perhaps 25 percent of total output—in the impoverished China of the 1930s. Janice
Perlman, who lived for some years among the poor people she was studying, in
The Myth of Marginality, Urban Poverty and Politics in Rio de Janeiro (Berkeley:
University of California Press, 1976), shows among other things that the shanty-
town dwellers of major cities in developing countries do not escape to the more
affluent parts of town over time but are condemned to permanent residence and
permanent deprivation in these squalid settlements. The almost unbelievable fact of
declining real incomes for the poor of India is documented in V. M. Dandekhar,
Poverty in India (Bombay: Indian School of Political Economy, 1971).

3. See Baran, *The Political Economy of Growth,* pp. 140–50, for the argument
and references.

4. See Keith Griffin, *The Political Economy of Agrarian Change, An Essay on
the Green Revolution* (Cambridge: Harvard University Press, 1974).

5. Oscar Lewis, in *Five Families* (New York: Basic Books, 1959),* offers a
poignant account of such fragmented lives at various levels in the income distribution in Mexico.

6. Richard J. Barnet and Ronald E. Miller in *Global Reach: The Power of
Multinational Corporations* (New York: Simon and Schuster, 1974), provide a useful
description and analysis of the ways in which multinational corporations inflict
social, political, and economic burdens on developing countries, while raking in
superprofits from the exploited citizenry.

7. See North American Congress on Latin America's publications, such as
*Yanqui Dollar, The Contribution of U.S. Private Investment to Underdevelopment
in Latin America* (Berkeley: NACLA, 1971).*

8. See Carl Riskin, "Surplus and Stagnation in Modern China," in Dwight
Perkins, ed., *China's Modern Economy in Historical Perspective.*

9. See notes 2 and 5 above. As an indication of the human costs of American
imperialism, one might note studies by the United Nations' Economic Commission
for Latin America and the El Salvador Ministry of Public Health (described in
La Prensa Grafica, El Salvador, Nos. 681204, 690122) which document the appalling
deprivation, including illiteracy and serious malnutrition, of some two-thirds of the
citizens of El Salvador.

Chapter 9

1. Very few American radicals are strong supporters of the Soviet Union these
days. However, there is a good deal of controversy surrounding the point in time
and the reasons that led the Russian revolution away from the path toward real
socialism. E. H. Carr's massive eight-volume *History of Soviet Russia* (Baltimore:
Pelican, 1952–76)* is generally regarded as providing the best account of the
crucial first fifteen years of Soviet socialism. Maurice Dobb's *Soviet Economic
Development Since 1917* (New York: International, 1948), is a very friendly account
of Soviet economic history. A much less favorable account appears in chapter 15,
vol. 2 of Ernest Mandel's *Marxist Economic Theory* (New York: Monthly Review,
1968).* Charles Bettelheim has embarked on a major reappraisal of the Russian
revolution, of which the first volume, *Class Struggles in the USSR* (New York:
Monthly Review, 1976), has recently appeared in English. The Yugoslav thesis that
the Soviet Union has become a state capitalist society, not fundamentally different
from its Western counterparts, is defended in Branko Horvat, *Toward a Theory of
Planned Economy* (Belgrade: Yugoslav Institute for Economic Research, 1968).

2. A useful survey containing various analyses of Soviet achievements and
failures can be found in the November 1967 issue of *Monthly Review,* which is devoted wholly to the reactions of a dozen leading scholars to the fiftieth anniversary
of the October Revolution.

3. For a short survey of twentieth-century European economic planning, see
my paper of that title in vol. 5, part II of the *Fontana Economic History of Europe,*
pp. 698–738, edited by Carlo Cipolla (London: Collins, 1976).

4. Most Americans and Europeans have been exposed to the massive campaign
associated with the publication of Solzhenitsyn's *Gulag Archipelago.* The many
horror stories detailed in that passionate account of the Soviet labor camps often

leave the impression that all Soviet citizens are slaves. Before accepting that idea, the reader is urged to compare the situation of the average Soviet citizen, or citizen of Eastern Europe, with his counterpart in developing capitalist countries. These Soviet "slaves" have a health status unmatched in that other world, have secure incomes and assured access to the basics of food, shelter, and health care. Their children are being well educated, with illiteracy virtually a thing of the past. Even those in the remote countryside have substantial access to modern cultural events; and the prospects for their children are very bright. As a final point, one might note that a study by a team of Soviet and East European economists showed that the Soviet Union's standard of living was among the lowest of the East European socialist countries, and that in particular in the mid-sixties it was less than two-thirds as high as in East Germany or Czechoslovakia. Clearly, this is not the same sort of thing as capitalist imperialism.

Chapter 10

1. The most important thing to understand about China is the revolution itself, its basis and its thrust. This story is best told in the small in the tales of transformation of individual villages. William Hinton's *Fanshen* (New York: Random House, 1966) * and Jan Myrdal's *Report from a Chinese Village* (New York: Random House, 1965) * are detailed and moving accounts of this process. There is no one book that describes well the economic history of revolutionary China. However, basic policy and organizational issues are discussed in Franz Schurmann, *Ideology and Organization in Communist China*, 2nd ed. (Berkeley: University of California Press, 1968). John G. Gurley, *China's Economy and the Maoist Strategy* (New York: Monthly Review, 1976), describes the main features of the Maoist approach to development and of the Chinese achievement.

Yugoslavia has proved to be a difficult country for Western socialists to make up their minds about. On the one hand, it has carried the principle of workers' management of factories farther than any other country, socialist or otherwise. On the other hand, it unabashedly uses markets to mediate relations among these worker-managed firms. Perhaps most radicals have initially reacted negatively to Yugoslavia, feeling that market relations were bound in the end to subvert the beneficent tendencies built into economic participation. However, the history of other socialist countries has seemed increasingly to suggest that some measure of decentralization of the economic process is really essential; and some form of marketlike relations is about the only alternative immediately available. Perhaps typical of this changing attitude is the exchange between two of the world's leading Marxist economists, American Paul Sweezy and Frenchman Charles Bettelheim, published in the United States as *On The Transition to Socialism* (New York: Monthly Review, 1971).* Both authors seem to be agreed that for a long time during the transition, properly controlled markets will in all likelihood be an important feature of the socialist scene.

The best account of Yugoslavia can probably be found in the works of Yugoslavia's own leading economist, Branko Horvat. His *Essay on Yugoslav Society* (New York: IASP, 1969) describes the issues that have faced Yugoslav socialists in more abstract terms.

2. The most useful account of Yugoslav policy and performance is still Branko Horvat, "Yugoslav Economic Policy in the Post-War Period: Problems, Ideas, Institutional Developments," *American Economic Review* 61 (June 1971), Supplement.

3. For a good account of the development of Maoist ideas during the course of the thirty-year revolution, see Gurley, *China's Economy and the Maoist Strategy*, esp. chap. 2.

4. For a stirring account of what it is like to live on a commune in China, see Jack Chen, *A Year in Upper Felicity* (New York: Macmillan, 1973).

5. For an analysis of industrial democracy, in general and in Yugoslavia, see Carole Pateman, *Participation and Democratic Theory* (Cambridge: Cambridge University Press, 1970).

6. For a plausible Maoist account, see Gurley, *China's Economy and the Maoist Strategy*, esp. chap. 1.

7. For a collection of readings on this topic, see Branko Horvat, Rudi Supek, and Mihailo Marković, eds., *Self-Governing Socialism, A Reader*, 2 vols. (White Plains, N.Y.: IASP, 1975).*

Chapter 11

1. Transitions are not easy to write about in general terms, as each country and each crisis has unique features that are central to understanding revolutionary potential. However, *Root and Branch, The Rise of the Workers' Movements* (Greenwich: Fawcett, 1975),* is a useful account of one of the central forces at work in capitalist society that is pushing toward the transition. Michael P. Lerner's *The New Socialist Revolution, An Introduction to Its Theory and Strategy* (New York: Delta, 1973),* is one attempt at such an account. Gar Alperovitz's "Socialism as a Pluralist Commonwealth," reprinted in Richard Edwards, Michael Reich, and Thomas Weisskopf, eds., *The Capitalist System, A Radical Analysis of American Society,* 2nd ed. (Englewood Cliffs, N.J.: Prentice-Hall, 1978),* pp. 522–39, describes a socialist society in terms of goals widely shared on the American left. The best way to study transitions, however, is to read accounts of successful and unsuccessful revolutions. Probably the most useful account of the origins and development of the Cuban revolution is Hugh Thomas's *Cuba* (New York: Harper and Row, 1971). Some useful accounts of Chile can be found in Ann Zammit, ed., *The Chilean Road to Socialism* (Sussex, England: International Development Studies, 1973).* Readings on Russia and China and Yugoslavia are noted in the last two chapters, though a full and friendly account of the Yugoslav revolution has not yet appeared in English.

Chapter 12

1. For Marx, see the *Communist Manifesto* itself and the *Critique of the Gotha Program,* for Lenin, *State and Revolution.*
2. Utopian novels often provide insight into the potential inhering in human nature, though of course they should not be taken as prescriptions for the future. Some of the more interesting works in this genre are Edward Bellamy's *Looking Backward* (Cambridge: Harvard University Press, 1967, orig. pub. 1888); William Morris, *News from Nowhere* (London: Reeves and Turner, 1890); and B. F. Skinner, *Walden Two* (New York: Macmillan, 1968).
3. Once again China provides an example of the thrust that socialism provides for more humane solutions. Though unable as yet to put substantial resources into massive pollution-control technology, the dispersion of factories and the massive development of small-scale industry have constituted important antipollution measures. And of course such measures are very closely associated with effective community building. For a good account of Chinese achievements in this area, see the American Rural Small-Scale Industry Delegation, *Rural Small-Scale Industry in the People's Republic of China* (Berkeley: University of California Press, 1977).

Chapter 13

1. As was noted in note 8 of chapter 8, Carl Riskin has made an estimate of the relative surplus for China in the thirties, building on work initiated by Victor Lippitt. But this work has not yet been extended to contemporary Third World countries.
2. Assar Lindbeck, *Political Economy of the New Left,* 2nd ed. (Harper and Row, 1977), pp. 75–77.

Chapter 14

1. Baran, *The Political Economy of Growth.* (New York: Monthly Review, 1957). The revised 1962 edition contains a foreword that is an extended commentary on the reviews of the 1957 edition.
2. Baran and Sweezy, *Monopoly Capital.* (New York: Monthly Review, 1966).
3. The surplus concepts are discussed in chap. 2 of *The Political Economy of Growth.*
4. Baran, *The Political Economy of Growth,* chap. 3.
5. Ibid., p. 119.

6. Ibid., pp. 124, 129.
7. Ibid., p. 133n.
8. Ibid.
9. This classification is discussd in chap. 6 and at the beginning of chap. 7 in Baran, *The Political Economy of Growth*.
10. Ibid., p. 257.
11. Ibid., 1962 ed., p. xxxvi.
12. The case is argued strongly in ibid., chap. 2, and has been repeated by Sweezy in a number of public lectures, and in a negative reaction to the Fitch-Oppenheimer finance-capital thesis in *Socialist Revolution*, (San Francisco: no. 8, 1972).
13. The case is argued in *Monopoly Capital*, chap. 4 and 8.
14. See chapter 5 above and note 2 to that chapter for this calculation.
15. Baran and Sweezy, in *Monopoly Capital*, pp. 135–38.
16. Ibid., p. 367.
17. Paul Sweezy and Charles Bettelheim, *On the Transition to Socialism* (New York: Monthly Review, 1971).
18. See chap. 11 of Leo Huberman and Paul Sweezy, *Socialism in Cuba* (New York: Monthly Review, 1969).
19. Harry Braverman, *Labor and Monopoly Capital, The Degradation of Work, in the Twentieth Century* (New York: Monthly Review, 1974).
20. Ibid., p. 180.

Chapter 15

1. This is the thesis of J. Steindl, *Maturity and Stagnation in American Capitalism* (Oxford: Oxford University Press, 1952). In his later work Baran emphasized the dialectical interaction between the two factors, the rate of investment and the rate of innovation, rather than attributing one-way causation from the latter to the former. However, the recent increase in popularity of the idea of the Kondratiev cycle, or long swing, has tended to reemphasize that line of causation in generating longer-term booms. For an example of this type of analysis, see Robert Zevin, "The Political Economy of the American Empire, December, 1974," in David Mermelstein, ed., *The Economic Crisis Reader* (New York: Random House, 1975).
2. A good, brief account of the historical volatility of capitalism that broadly fits our optimal radical interpretation is Eric Hobsbawm, "Capitalist Crises in Historical Perspective," in Richard Edwards, Michael Reich, and Thomas Weisskopf, *The Capitalist System, A Radical Analysis of American Society*, 2nd ed. (Englewood Cliffs, N.J.: Prentice-Hall, 1978), pp. 431–40.
3. Stagnation is at center stage in Ernest Mandel's *Marxist Economic Theory* (New York: Monthly Review, 1968) (Fr. ed. 1962), esp. chap. 14. In his more recent *Late Capitalism* (New York: Monthly Review, 1976), Mandel's views are somewhat equivocal but can still reasonably be described as stagnationist.
4. A good sampling of relevant radical writings reflecting these factors can be found in Mermelstein, *The Economic Crisis Reader* in *Radical Perspectives on the Economic Crisis of Monopoly Capitalism* (Union for Radical Political Economics, 1975),* and *U.S. Capitalism in Crisis* (New York: Union for Radical Political Economics, 1978).*
5. Perhaps most notably by Paul Hoffman, former chief executive of Studebaker, in a series of speeches in the fifties, after he turned public "servant."
6. Theresa Hayter's *Aid as Imperialism* (Baltimore: Penguin, 1971), provides a more wholesome account on this aspect of exploitation.
7. A crude reproduction of the calculation procedure used by Joseph Phillips in the Appendix he wrote to *Monopoly Capital*, based on data taken from the *Economic Report of the President* for 1976 and for 1973, gives the following: From 1963 to 1973 surplus as a share of national income increased from 67.8 to 68.7 percent. Surplus absorbed by government increased from 34.7 to 38.0 percent of national income. National defense expenditures increased by a little over 50 percent in the decade, but as a share of gross national product declined from 8.8 to 5.8 percent. Thus the increased relative surplus absorption is accounted for by nondefense government activity. Notable in this latter area is a more than tenfold increase in federal health expenditures and a more than trebling of total government expenditures on education and manpower training. The relevant data can be found in the 1976 *Economic Report*, (Washington, D.C.: U.S. Government Printing Office) tables B-11, B-12, B-64, and B-69.

Chapter 16

1. Karl Marx, *Early Writings* (New York: McGraw-Hill, 1963).*
2. For an analysis of Marx's views, see Bertell Ollman, *Alienation, Marx's Conception of Man in Capitalist Society,* (Cambridge: Cambridge University Press, 1971).* A useful if non-Marxist account of the problem under contemporary capitalism is Walter A. Weisskopf, *Alienation and Economics* (New York: Dutton, 1971). For the "Marxist-Humanist" perspective, see Howard Sherman's *Radical Political Economy* (New York: Basic Books, 1972), esp. chap. 11 and 21 and references.
3. Herbert Marcuse, *One-Dimensional Man* (Boston: Beacon, 1964),* esp. chapters 1–4, 10. The concepts in quotations in the next few paragraphs in the text are all taken from those chapters.
4. The last two chapters of Branko Horvat's *Essay on Yugoslav Society* (New York: IASP, 1969) contain some relevant arguments. A selection of readings edited by Horvat, the Yugoslav philosopher Rudi Supek, and Mihailo Marković (White Plains, New York: IASP, 1975) * *Self-Governing Socialism, A Reader,* 2 vols., contains a number of selections relevant for alienation and disalienation, several of which are by Yugoslavs. See vol. 1, pp. 327–50, 363–65, 405–37.

Chapter 17

1. Several of Horvat's works have been cited in chapters 1, 10, and 16. The present chapter is a revised and shortened version of my "Marxism-Horvatism, A Yugoslav Theory of Socialism," *American Economic Review* 57 (June 1967).
2. The issue of workers' control has been much discussed in recent years and interest in the topic seems to be still rising. In addition to the already cited book of readings edited by Horvat, Supek, and Marković, there are two other interesting collections: Gerry Hunnius et al., eds., *Workers' Control* (New York: Vintage, 1973),* and Jaroslav Vanek, ed., *Self-Management* (Baltimore: Penguin, 1975).*
3. Issues related to the efficiency of worker-managed enterprises are appraised by Vanek in his *General Theory of the Labor-Managed Market Economy* (Ithaca, N.Y.: Cornell University Press, 1969). Vanek and Horvat both believe that such enterprises will prove to be very efficient; for a less positive view, see my *Socialist Economy, A Study of Organizational Alternatives* (New York: Random House, 1967), chap. 8–10.
4. Paul Sweezy and Charles Bettelheim, *On the Transition to Socialism* (New York: Monthly Review, 1971).*

Chapter 18

1. This elitist character of neoclassical economics is described and analyzed in my *What's Wrong With Economics* (New York: Basic Books, 1972), esp. Parts I and III–IV.
2. The standard work of contemporary mathematical Marxism is Michio Morishima, *Marx's Economics* (Cambridge: Cambridge University Press, 1973). For an alternative and perhaps more nearly Marxist though still technical formulation of the issue of discrimination, see John Roemer, "Differentially Exploited Labor: A Marxian Value Theory of Discrimination," mimeo, 1976; this and other of Roemer's papers show this genre at its most productive. The controversy with neoclassical economists over capital and the theory of growth is described in G. C. Harcourt, *Some Cambridge Controversies in the Theory of Capital* (Cambridge, England: Cambridge University Press, 1972).* A very good introduction to economic theory from a radical but not really Marxist perspective is Joan Robinson and John Eatwell, *An Introduction to Modern Economics* (London: McGraw-Hill, 1973).* For an application of value theory to international trade, see Arghiri Emmanuel, *Unequal Exchange* (New York: Monthly Review, 1972).

One of the most interesting of recent controversies among radical economists concerns the role of financial as opposed to industrial interests in controlling monopoly capitalist economies. The discussion has taken place principally in the pages of *Socialist Revolution*, issues 4–6 (1970–71), 8 (1972), 11–12 (1974), and the main protagonists are Robert Fitch and Mary Oppenheimer, James O'Connor, Paul Sweezy, and others. An empirical appraisal of finance power has recently been completed by David Kotz, *The Role of Financial Institutions in the Control of Large Nonfinancial Corporations* (Berkeley: University of California Press, 1977). Other controversies, such as that over the prospects for continued cooperation among leading capitalist countries in international affairs or the importance of political factors in the business cycle, have flared up in recent years and been discussed constructively, offering clear evidence that radical economic scholarship is now established on a firm base of agreement on most fundamentals.

Suggestions for Further Reading*

Baran, Paul. *The Political Economy of Growth.* New York: Monthly Review, 1957, rev. ed. 1962.*
    Probably the single most important Marxist work of the postwar era, it must be understood and appraised by every serious student of radical economics.
Baran, Paul, and Paul Sweezy. *Monopoly Capital.* New York: Monthly Review, 1966.*
    The radical textbook in the United States in the later sixties, it is still a useful guide to the "surplus-absorption" theory of capitalist crisis.
Barnet, Richard J., and Ronald E. Miller. *Global Reach, The Power of the Multinational Corporations.* New York: Simon and Schuster, 1974.
    The best book on this difficult and controversial subject.
Bowles, Samuel, and Herbert Gintis. *Schooling in Capitalist America.* New York: Basic Books, 1976.
    A powerful and scholarly account of the effect of class on American education.
Braverman, Harry. *Labor and Monopoly Capital, The Degradation of Work in the Twentieth Century.* New York: Monthly Review, 1974.
    The one best book on this central topic.
Brown, Michael Barratt. *The Economics of Imperialism.* Baltimore: Penguin, 1974.*
    A solid, modern, Marxist analysis of the issue.
Dobb, Maurice. *Studies in the Development of Capitalism,* rev. ed. New York: International, 1963.*
    A key work in economic history by one of the great radical economists of the era of relative stagnation in radical economic research (1920-55).
Dowd, Douglas. *The Twisted Dream, Capitalist Development in the United States since 1776,* 2nd ed. Cambridge, Mass.: Winthrop, 1977.*
    A passionate and useful account.
Eaton, John. *Political Economy.* International, 1963.*
    One of the two best primers of the economics of Karl Marx.
Edwards, Richard, Michael Reich, and Thomas Weisskopf, eds. *The Capitalist System,* 2nd ed. Englewood Cliffs, N.J.: Prentice-Hall, 1978.*
    A collection of readings that comprise the best introductory textbook on radical political economy currently available.
Frank, Andre Gunder. *Capitalism and Underdevelopment in Latin America.* New York: Monthly Review, 1967.*
    In this and other works Frank has attempted to flesh out Baran's "creation of underdevelopment" thesis, using evidence, in this case, from Chile and Brazil.
Gordon, David, ed. *Problems in Political Economy: An Urban Perspective,* 2nd ed. Boston: Heath, 1977.*
    A very useful collection on political economic issues of labor, race, crime, poverty, education and health.
Gurley, John. *China's Economy and the Maoist Strategy.* New York: Monthly Review, 1976.
    A collection of essays by a leading defender of Maoist political economy.
Hinton, William. *Fanshen, A Documentary of Revolution in a Chinese Village.* New York: Random House, 1966.*
    Probably the one best place to get a grass-roots understanding of the meaning and process of revolution in the countryside.
Horvat, Branko. *An Essay on Yugoslav Society.* White Plains, N.Y.: International Arts and Sciences, 1969.
    Repeats some of the material in his 1964 work, but with several chapters more specifically devoted to Yugoslavia and its relation to Horvat's "optimal regime".
————. *Toward a Theory of Planned Economy.* Belgrade: Yugoslav Institute for Economic Research, 1964.

* Starred items are available in paperback edition.

As noted in the text, one of the half-dozen seminal works of postwar radical economics.

————; Rudi Supek; and Mihailo Markovic̓, eds. *Self-Governing Socialism, A Reader,* 2 vols. White Plains, N.Y.: International Arts and Sciences, 1975.*

A collection of readings covering the gamut of philosophical, social, and economic issues related to economic participation by the direct producers in the decisions that affect their lives.

Huberman, Leo, and Paul Sweezy. *Socialism in Cuba.* New York: Monthly Review, 1969.*

An account that shows what revolution can mean to the peoples of Latin America.

Marcuse, Herbert. *One-Dimensional Man.* New York: Beacon Press, 1964.*

Though not an economist, Marcuse's treatment of alienation is the most relevant of recent efforts for the political economist, though perhaps an overly pessimistic appraisal.

Mermelstein, David, ed. *The Economic Crisis Reader.* New York: Random House, 1975.*

After reading this book the reader will not doubt that we are in the midst of a great crisis.

Miliband, Ralph. *The State in Capitalist Society.* New York: Basic Books, 1969.

Why reformism fails as an instrument of radical change.

Morishima, Michio. *Marx's Economics.* Cambridge: Cambridge University Press, 1973.

The standard work of contemporary mathematical Marxism. Demanding reading but does not provide real insight into the working of contemporary monopoly capitalism.

O'Connor, James. *The Fiscal Crisis of the State.* New York: St. Martins, 1973.*

How the state, while trying to serve the needs of the capitalists, has boxed itself into a corner.

Poulantsas, Nicos. *Classes in Contemporary Capitalism.* London: New Left Books, 1975.

Along with Miliband and O'Connor, one of the key contemporary works on the role of classes and the state under monopoly capitalism.

Robinson, Joan, and John Eatwell. *An Introduction to Economics.* New York: McGraw-Hill, 1973.*

A difficult, unconventional, rewarding textbook.

Schurmann, Franz. *The Logic of World Power, An Inquiry into the Origins, Currents and Contradictions of World Politics.* New York: Random House, 1974.*

A challenging work in an area of still underdeveloped radical research.

Sherman, Howard. *Radical Political Economy.* New York: Basic Books, 1972.

The best account of the subject from the "Marxist-humanist" perspective.

Sweezy, Paul. *Theory of Capitalist Development.* New York: Monthly Review, 1942.*

One of the two best primers of the economics of Karl Marx.

Tuchman, Barbara. *The Proud Tower, A Portrait of the World Before the War: 1890–1914.* New York: Macmillan, 1966.*

One can feel the enveloping crisis of capitalism through daily lives in this gripping account.

Union for Radical Political Economics. *Radical Perspectives on the Economic Crisis of Monopoly Capitalism.* New York: URPE, 1975.*

————. *U.S. Capitalism in Crisis.* New York: URPE, 1978.

Together with the Mermelstein readings, these books provide a good perspective on the current crisis.

Vanek, Jaroslav. *General Theory of the Labor-Managed Market Economy.* Ithaca, N.Y.: Cornell University Press, 1969.

The economic theory of self-management by a believer in both.

Wallerstein, Immanuel. *The Modern World-System.* New York: Academic, 1974.

A challenging account of the rise of the capitalist mode of production and of the process by which it penetrated new areas.

Index

Agriculture: historical development of, 16–18; collective, 70–71
Algeria: overthrow of reactionary government in, 84, 85; social effects of colonialism in, 12
Alienation, 116–20; expenditures related to, 38
Allende, Salvatore, 82
Alperovitz, Gar, 143
Althusser, Jean, 135
American Rural Small-Scale Industry Delegation, 143
Anarchy of market, 54
Angola, Soviet support of, 56
Arms Control and Disarmament Agency, U.S., 136
Automobile industry, waste in, 36–37

Bangla Desh, capitalist institutions in, 57
Baran, Paul, 5, 35, 64, 98, 99, 102–11, 113–15, 124, 127, 135, 138–41, 143, 144; on political economy of growth, 103–6
Barnet, Richard J., 141
Bellamy, Edward, 143
Benassy, J. C., 140
Berg, Alan, 136
Bettelheim, Charles, 79, 126, 135, 141, 142, 144, 145
Blacks: wage rates for, 11; see also Racism
Block, Fred, 140
Bloomfield, Lincoln, 138
Boddy, Raford, 140
Bolivia: "growth fascism" in, 62; overthrow of reactionary government in, 84, 85; radicalism in, 4
Bolsheviks, 67, 71
Bottomore, T. B., 139
Bowles, Samuel, 139
Braverman, Harry, 109, 110, 138
Brazil: economic growth in, 64; "growth fascism" in, 62; United States imperialism and, 10
Briffault, Robert, 136
Britain: government regulation of business in, 26; Great Depression in, 35; imperialism of, 58, 60; industrial revolution in, 19, 20; intermittent crisis in, 55; nineteenth century, 22; occupation of Murmansk by, 29; rise of Labour party in, 23; worker control in, 122
Brown, E. H. Phelps, 137
Bukharin, Nikolai, 137
Bulgaria, life of average citizens in, 72
Bureau of the Census, U.S., 136
Bureaucracies: in Soviet Union, 88; in Third World countries, 123
Business Week, 104

Cadre-building, 85
Cambridge Journal of Economics, 5
Capital (Marx), 131
Carr, E. H., 141
Central Intelligence Agency (CIA), 62
Centralization, 46
Chad, capitalist institutions in, 57
Chen, Jack, 142
Chiang Kai-shek, 10
Childe, V. Gordon, 137
Chile: Allende government in, 82–83; capitalist institutions in, 57; counterrevolution in, 102; radicalism in, 4; United States imperialism and, 10
China: ancient, 16; British imperialism and, 58; cadre-building in, 85; collectivized agriculture in, 71; decentralization in, 79; economic development in, 76; economic education of peasants in, 130; English language publications from, 6; families in, 91; Nationalist, see Taiwan; participation in, 77; reversal of cultural revolution in, 88; socialist revolution in, 66, 84; Soviet conflict with, 107; surplus extraction in, 64; technology in, 92; transformation of man in, 80–81; vanguard-party thesis and, 125; violent revolution in, 102
CIA, 62
Cipolla, Carlo, 138, 141

Civil rights, 86
Class, 11; Marx on role of, 15
Class struggle: instability and, 52–53; urbanization and, 50–51
Cohen, C. D., 140
Cold War, 30
Collectivization, 70–71; in Yugoslavia, 75
Colonial regimes, 105
Colonialism, 12
Cominform, 75
Commerce Department, U.S., 138
Communes: in China, 76; in Yugoslavia, 78
Communications media, 51; access to, in United States, 86–87; development and, 60
Communist party of Soviet Union, 68
Communist party of Yugoslavia, 78
Competing capitals, 53–54
Comprador regimes, 105
Congress, U.S., domination by business interests of, 25
Consumption, per capita, 36
Crisis, 50–56; government, 47–49; monopoly capitalism as response to, 24–27; nineteenth century, 21–23
Crotty, J., 140
Cuba: elitism of leadership in, 88; Leninist party in, 85; publications on, 6; revolution in, 84; Soviet support of, 56, 70
Czechoslovakia: collective farming in, 71; Soviet suppression of revolution in, 72

Dandekhar, V. M., 141
Debt peonage, 18, 19
Demand, volatility of, 51
Developing countries, *see* Development
Development: imperialism and, 57–65; uneven, 63–65
Discrimination, 11
Dobb, Maurice, 127, 128, 137, 141
Domhoff, G. William, 139
Dominican Republic, United States intervention in, 10, 62, 63
Dow, J. C. R., 140
Dowd, Douglas, 137, 140
Drucker, Peter, 140
Drug industry, waste in, 36–37

Eaton, John, 136
Eatwell, John, 145
Ecological problems, 264

Economic instability, nineteenth century, 22
Education, government expenditures for, 44
Edwards, Richard, 136, 138, 143
Egypt: ancient, 16; overthrow of reactionary government in, 84, 85
El Salvador Ministry of Public Health, 141
Emmanuel, Arghiri, 145
Engels, Friedrich, 137
Eurocommunism, 83
Exegetics, Marxist, 131
Exploitation under monopoly capitalism, 33–41

Factory system, development of, 20
Family, social change and, 91
Fannon, Franz, 136
Fascism, 29, 35; "growth," 62
Field, Alexander, 139
Financial markets: development of, 22; volatility of, 51
Fitch, Robert, 145
Flynn, Dennis O., 139
Ford Motor Company, 36
Foreign aid, 62
Foreign investment, 61–62
France: domestic communism in, 30; imperialism of, 60; inflation in, 55; population of, 79; transition to socialism in, 83
Frank, Andre G., 140
Franklin, Raymond, 136
French Revolution, 4

Germany: counterrevolution in, 102; government regulation of business in, 25–26; revolutionary struggles in, 29; social democracy in, 23, 83; worker control in, 122; in World War II, 30; *see also* West Germany
Gintis, Herbert, 139
Goldberg, Marilyn, 138
Gordon, David, 138, 139
Government: alliance of business and, 31; inefficient, 51–52; monopoly capitalism and, 42–49; normal and crisis, 47–49
Great Depression, 29, 34–35
Greece: ancient, 17; capitalist institutions in, 57; civil war in, 30; "growth fascism" in, 62; United States imperialism and, 10
Green, Mark J., 139
Green Revolution, 60

Griffin, Keith, 141
Gross, Bertram, 138
Growth: political economy of, 103–6; surplus extraction and, 35–39
"Growth fascism," 62
Guevara, Ché, 6
Gurley, John G., 136, 142

Habermas, Jürgen, 135
Handfield-Jones, 137
Harcourt, G. C., 145
Harris, Seymour, 138
Hayter, Theresa, 144
Hegel, Georg Friedrich Wilhelm, 117
Hinton, William, 136, 142
Hobsbawm, Eric, 137
Hoffman, Paul, 144
Honduras, capitalist institutions in, 57
Horvat, Branko, 5, 6, 121–26, 128, 135, 137, 141, 142, 145
Huberman, Leo, 144
Hungary: decentralization in, 78; factory committees in, 122; Soviet suppression of revolution in, 72
Hunnius, Gerry, 145

Ibsen, Henrik, 136
Illegal activities, 37
Imperialism, 9–10; development and, 57–65; rise of, 22; of Soviet Union, 72
Income distribution, 37–38; discrimination and, 183
India: ancient, 16; British imperialism in, 28; capitalist institutions in, 57; increasing poverty in, 64
Industrial regulation, involvement of business interests in, 25–26
Industrial revolution, 19–21; British imperialism and, 58; surplus extraction and, 34
Industrial Workers of the World (IWW), 23
Inequality, 9–10
Insecurity of modern life, 39–40
Instability, 50–56; crisis and, 55–56; in nineteenth century, 22; tendencies promoting, 50–52; underlying forces of, 52–55
Internationalization of capitalism: instability and, 51; during nineteenth century, 22
Interstate Commerce Commission (ICC), 25
Investment, foreign, 61–62

Iraq, overthrow of reactionary government in, 84
Italy: domestic communism in, 30; transition to socialism in, 83
ITT, 83

Jalée, Pierre, 138
Japan: economic growth in, 31; foreign investments of, 31; industrialization of, 57, 105; United States boycott of, 29; in World War II, 30

Kershaw, Joseph, 308
Keynes, John Maynard, 54
Keynesianism, 26, 30, 140; taming of business cycle and, 54–55
Kissinger, Henry, 83
Kolko, Gabriel, 137
Kolko, Joyce, 140
Kondratiev cycle, 144
Korea, United States imperialism and, 10
Korean War, 55
Kotz, David, 139, 146
Kuznets, Simon, 136

Labor and Monopoly Capital, The Degradation of Work in the Twentieth Century (Braverman), 109
Labor struggles: during nineteenth century, 23; during twentieth century, 26
Labour party of Britain, 23
Late Capitalism (Mandel), 5
Law of value, 34
Lebanon: United States imperialism and, 10; United States intervention in, 62
Leijonhufvud, Axel, 140
Leiss, Amelia, 138
Lenin, V. I., 68, 71, 90, 137, 143
Leninist parties, 84–85, 89
Lerner, Michael P., 143
Lewis, Oscar, 141
Lindbeck, Assar, 100, 143
Lippitt, Victor, 143
Liu Shao Chi, 137
Loss-producing economic activity, 37
Lyons, Nancy, 136

Malnutrition, 9
Malraux, André, 136

Mandel, Ernest, 5, 128, 135, 137, 141
Mantoux, Paul, 138
Mao Tse-tung, 5, 6, 135, 137
Maoism, 76
Marcuse, Herbert, 63, 128, 136, 145; on alienation, 117–20
Marglin, Stephen, 137
Market: anarchy of, 54; financial, 22, 51
Marković, Mihailo, 142, 145
Marx, Karl, 15–16, 90, 121, 127, 131, 134, 136–37, 143, 145; on alienation, 117; theory of exploitation of, 40–41
Marxism, 97; alienation and, 117; exegetic, 131; mathematical, 130; modifications of, 122; neoclassical economics and, 129; recent works based on, 5; social democrats and, 83; on technology, 91, 92
Marxism-Leninism, 137
Marxist Economic Theory (Mandel), 5
Mass media, *see* Communications media
Mathematical Marxism, 130
Meiji Restoration, 105
Mermelstein, David, 138, 144
Mesopotamia, 16, 17
Militarization, 11; Pax Americana and, 31–32
Military expenditures, 42
Miller, Herman P., 136
Miller, Ronald E., 141
Mises, Ludwig von, 122
Monopoly Capital (Baran and Sweezy), 5, 102, 107, 108, 115
Monopoly capitalism, 8–14; alienation and, 11–13; exploitation under, 205–13; family and, 91; government and, 42–49; growth of surplus under, 35–39; imperialism and, 10–11; inequality of, 9–10; insecurity of life under, 39–40; irrationality of, 13; racism and sexism under, 11; revolutionary process and, 86–88; structure and tendencies of, 24–32; underdevelopment and, 57
Monthly Review, 5
Morgan, J. P., 24
Morishima, Michio, 145
Morocco, capitalist institutions in, 57
Morris, William, 90, 143
Moss, Stanely, 138
Multinational corporations, 114
Murmansk, British occupation of, 29
Myrdal, Jan, 136, 142

Nader, Ralph, 139
Nazis, 71

Neoclassical economics, 129–31
Netherlands, inflation in, 55
New Deal regimes, 105, 106
New Left, 6, 109
New Left Review, 5
New York City, financial crisis in, 48
News from Nowhere (Morris), 90
Nexö, Martin Anderson, 136
Nicaragua, life of average citizens in, 72
Nineteenth century capitalism, crisis of, 21–23
Nixon, Richard: economic mismanagement under, 52; fall of, 46
Nuclear family, 91
Nuclear war, 86

Occupations, wasteful, 37
O'Connor, James, 139, 145
October Revolution, 90
Old Left, 6, 109
Ollman, Bertel, 145
OPEC, 53
Opium War, 58
Oppenheimer, Mary, 145
Output per capita, 35

Paris Commune, 67
Participation, socialism and, 77–78
Pateman, Carole, 142
Pax Americana, 26, 31
Peasantry, development of, 16–18
Perlman, Janice, 141
Perlo, Victor, 136, 140
Peru, overthrow of reactionary government in, 84
Phillips, Joseph, 139, 144
Poland, factory committees in, 122
Political Economy (journal), 6
Political Economy of Growth (Baran), 5, 102, 107
Population distribution, 21
Portugal, imperialism of, 60
Portuguese Africa, 84
Poverty, inequality and, 9
Prats, General, 82
Preobrazhensky, Evgeniy, 137
Production: forces and relations of, 20; socialization of, 46

Racism, 11
Radical scholarship, 128–29
Railroads, 22
Real wages, stagnation of, 38
Regulation of industry, involvement of business interests in, 25–26
Reich, Michael, 136, 138, 143

Resnick, Solomon, 136
Review of Radical Political Economy (journal), 5
Revolutions, 79–81, 84–86; nineteenth century, 21; role of, 132–34
Riskin, Carl, 141
Rivalry among capitalists, 21–22
Robinson, Joan, 145
Roemer, John, 145
Rome, ancient, 17
Russia, *see* Soviet Union
Russian Revolution, 66–68, 79, 122

Sampson, Anthony, 140
Schmidbauer, P. L., 138
Scholarship, radical, 128–29
Schurmann, Frank, 136, 142
Serfdom, 17
Sexism, 11
Share tenancy, 18
Shelly, Alan, 140
Sherman, Howard, 140, 145
Shoshone Indians, 33
Siberia, United States occupation of, 29
Silk, Leonard, 139
Skinner, B. F., 143
Slavic family structures, 91
Smith, Adam, 54
Social democracy, 83–84
Social-harmony expenditures, 44–46
Socialism: defense against capitalist depradations of, 4; rise of, 66–81; size and power of movements for, 4; in Soviet union, 66–73; transition from capitalism to, 82–89
Socialist party of United States, 23
Socialist Revolution (journal), 5
Socialization of production, 46
Solzhenitsyn, Alexander, 141
Soviet Union, 29, 66–73; achievements of socialism in, 68–70; Chinese conflict with, 107; in Cold War, 30; Communist party of, 5–6; failures of, 70–72; oppressive bureaucracy in, 88; pollution in, 92; support of national liberation struggles by, 56; suppression of dissent in, 101; in World War II, 30
Stagflation, increasing tendency toward, 52
Stagnation, surplus absorption and, 112–15
Stalin, Josef, 71
Steinbeck, John, 136
Steindl, J., 107, 144
Sugar Trust, 24
Supek, Rudi, 142, 145
Supreme Court, U.S., 30

Surplus absorption, 112–15
Surplus extraction, 33–35; in developing nations, 63; growth of, 35–39; insecurity of modern life and, 39–40
Sweden, inflation in, 55
Sweezy, Paul, 5, 98, 99, 102–11, 113–15, 124, 126–28, 135, 137–40, 142–45
Syria, overthrow of reactionary government in, 84, 85
Szechuan province, population of, 79

Taiwan: capitalist institutions in, 57; "growth fascism" in, 62; United States imperialism and, 10
Technology: centralization and, 46; development and, 60; foreign investment and, 61–62; industrial revolution and, 20; military, 19, 20; socialism and, 91–92
Thieu, General, 10
Third World: bureaucracies in, 123; capitalist institutions in, 57; decolonization of, 30, 55; delegitimization of United States policies toward, 87; economists from, 6; influence of Communist parties in, 70; living conditions in, 64; losses imposed by imperialism on, 58–59; multinational corporations in, 114; resource constraints and, 113; social development processes and, 60; surplus extraction in, 98–99
Thomas, Hugh, 143
Tobacco Trust, 24
Toward a Theory of Planned Economy (Horvat), 5
Trade, development and, 61
Transformation of man, 80–81
Transportation, 51; development and, 60
Trust-busting, 25
Tuchman, Barbara, 137

Ukraine, collective farms in, 71
Underdeveloped nations, *see* Development; Underdevelopment
Underdevelopment, continuance of, 59–63; creation of, 57–59
Uneven development, 63–65
Union of Soviet Socialist Republics, *see* Soviet Union
United Kingdom, *see* Britain
United Nations: Economic Commission for Latin America of, 141; Food and Agriculture Organization of, 136

United States: access to communications media in, 86–87; alienation in, 12; consumption per capita in, 36; decline of radicalism in, 4; economic aid from, 30; government expenditures in, 42–49; government regulation of business in, 25; growth of output per capita in, 35; hegemony of, 55; imperialism and, 10, 60; industrial revolution in, 19, 105; labor struggles in, 23, 26–27; nineteenth century capitalism in, 22; nuclear capability of, 86; occupation of Siberia by, 29; opposition to Allende government by, 83; pollution in, 92; potential use of force by, 103; poverty in, 9; worker control in, 122

Upton, Letitia, 136

Urbanization, intensification of class struggle and, 50–51

U.S.S.R., *see* Soviet Union

U.S. Steel, emergence of, 24

Utopian societies, 90–91

Value, law of, 34

Vanke, Jaroslav, 145

Venceremos (Gerassi), 6

Vietnam: overthrow of reactionary government in, 84; Soviet support of, 56; United States imperialism and, 10

Vietnam War, 55, 63

Vogel, David, 139

Volatility of demand, 51

Wage labor, generalized system of, 34

Wang, George, 135

War, expenditures for, 43–44

Waste: in occupations, 37; in production, 36–37

Watergate, 48

Wealth, ownership of, 9

Weber, Max, 122

Weimar Germany, 83

Weinstein, James, 137

Weisskopf, Thomas, 136, 138, 143

Weisskopf, Walter A., 145

Welfare, government expenditures for, 44–45

West Germany: foreign investments of, 31; as hard-currency nation, 55; population of, 79

William, William A., 136

Women, wage rates for, 11

Worker control, 122

Working class, organized, 26; *see also* Labor struggles

World War I: causes of, 27; economic controls during, 26; full employment during, 35; military expenditures during, 42

World War II, 108, 138; changes in American economy during, 35; military expenditures during, 42; monopoly capitalism and, 29–30

Youth movement, radical, 6

Yugoslavia: aid to, 10; alienation and, 118–19; cadre-building in, 85; commercialism in, 88; decentralization in, 78–80; economic development in, 75–76; participation in, 77–78; publications on, 6; socialist revolution in, 66, 84; transformation of man in, 80; worker control in, 122

Zammit, Ann, 143

Zevin, Robert, 144